An Introduction to Reactive Power Control and Voltage Stability in Power Transmission Systems

ABHIJIT CHAKRABARTI
*Professor
Department of Electrical Engineering
Indian Institute of Engineering Science and Technology (IIEST)
Shibpur, Howrah
Former Vice Chancellor, Jadavpur University, Kolkata*

D.P. KOTHARI
*Former Vice Chancellor
VIT University, Vellore
and Director-in-Charge, IIT Delhi*

A.K. MUKHOPADHYAY
*Former Vice Chancellor, Tripura University
and Professor of Electrical Engineering
Department of Applied Physics
University of Calcutta*

ABHINANDAN DE
*Assistant Professor
Department of Electrical Engineering
Indian Institute of Engineering Science and Technology (IIEST)
Shibpur, Howrah*

PHI Learning Private Limited

Delhi-110092
2015

"Dedicated to Our Parents"
—Authors

₹ 250.00

AN INTRODUCTION TO REACTIVE POWER CONTROL AND VOLTAGE STABILITY IN POWER TRANSMISSION SYSTEMS
A. Chakrabarti, D.P. Kothari, A.K. Mukhopadhyay and Abhinandan De

© 2010 by PHI Learning Private Limited, Delhi. All rights reserved. No part of this book may be reproduced in any form, by mimeograph or any other means, without permission in writing from the publisher.

ISBN-978-81-203-4050-3

The export rights of this book are vested solely with the publisher.

Third Printing   ...   .   **June, 2015**

Published by Asoke K. Ghosh, PHI Learning Private Limited, Rimjhim House, 111, Patparganj Industrial Estate, Delhi-110092 and Printed by Raj Press, New Delhi-110012.

Contents

Preface   *ix*

1. **POWER TRANSMISSION IN UNCOMPENSATED AC TRANSMISSION LINES**   1

   1.1 Introduction   *1*
   1.2 Electrical Parameters of Transmission Lines   *1*
   1.3 Representation of a Transmission Line and Transmission Line Equation   *2*
   1.4 Concept of Power in AC Transmission Systems   *5*
   1.5 Power Flow in a Two-terminal Power Transmission Network   *7*
   1.6 Receiving-end Power Circle Diagram   *11*
   1.7 Reactive Power Requirement of an Uncompensated Line   *14*
   1.8 Implication of Surge Impedance Loading   *16*
   1.9 Reactive Loss Characteristics of Transmission Lines   *17*
   1.10 Operation of a Transmission Line at No-Load Condition   *19*
   1.11 Operation of a Transmission Line under Heavy Loading Condition   *24*
   1.12 Voltage Regulation of the Transmission Line and its Relation with Reactive Power   *26*
   1.13 Maximum Power Transfer in an Uncompensated Line   *32*
   1.14 Line Loadability   *35*
   Exercises   *38*

2. **REACTIVE POWER FLOW AND VOLTAGE CONTROL PROBLEMS**   39

   2.1 Introduction   *39*
   2.2 Reactive Power-Voltage (Q-V) Coupling Concept   *39*
   2.3 Governing Effects on Reactive Power Flow   *41*
       2.3.1 Line Demand   *41*
       2.3.2 Effect of Receiving-end Bus Loading   *41*
       2.3.3 Implication of Voltage Regulation   *42*
       2.3.4 Implication of Magnitude of Power Angle   *42*
       2.3.5 Load Bus Reactive Power Sensitivity   *43*
       2.3.6 Effect of Series Reactive Loss   *43*

2.4 Aspects of Real and Reactive Power Static Stability   46
    2.4.1 Steady State (Static) Real Power Stability   46
    2.4.2 Steady State (Static) Reactive Power Stability   47
2.5 Real and Reactive Power Transient Stability   49
    2.5.1 Real Power Transient Stability   50
    2.5.2 Reactive Power Transient Stability   51
2.6 Concept of Dynamic Stability   52
2.7 Relation between Voltage and Reactive Power at a Node in a Power System   53
2.8 Reactive Power Requirement for Control of Voltage in Long Lines   54
2.9 Operational Aspects in Reactive Power and Voltage Control   55
    2.9.1 System MVAR Mismatch   56
    2.9.2 Vulnerable System Disturbance   57
2.10 Basic Principle of System Voltage Control   58
2.11 Reactive Power Flow Constraints and their Implications in Loss of Voltage   58
2.12 Effect of Transformer Tap Changing in the Post-disturbance Period   60
2.13 Effect of Generator Excitation Adjustment in the Post-disturbance Period   61
2.14 Practical Aspects of Reactive Power Flow Problems Leading to Voltage Collapse in EHV Lines   61
Exercises   63

3. VOLTAGE STABILITY—FUNDAMENTAL CONCEPTS   64

3.1 Introduction   64
3.2 Reactive Power and Voltage Collapse   64
3.3 Changes in Power System Contributing to Voltage Collapse   65
3.4 Concept of Stability of Transmission System   65
3.5 Relation of Voltage Stability to Rotor Angle Stability   66
3.6 Stability Margins   66
3.7 Definition and Classification of Voltage Stability   70
    3.7.1 Definition of Voltage Stability, Voltage Instability and Voltage Collapse   70
    3.7.2 Classification of Power System Stability   71
3.8 Mechanism of Voltage Collapse   72
3.9 Analysis of Power System Voltage Stability   74
    3.9.1 A Simple Example   74
    3.9.2 Analysis of Voltage Stability of Non-linear Power System   75
    3.9.3 Factors Affecting Voltage Stability   77
3.10 Voltage Collapse and Modelling of Voltage Collapse   89
    3.10.1 Modelling of Voltage Collapse   89

3.11 Real and Reactive Power Operating Contour of a Radial
Power Transmission System   90
3.12 Basic Aspects of Voltage Stability   *91*
3.13 Voltage Security   *93*
3.14 Transient Voltage Stability   *95*

Exercises   *99*

4. POWER TRANSFER AT VOLTAGE STABILITY LIMIT   100

4.1 Introduction   *100*
4.2 Magnitude of Receiving-end Voltage   *100*
4.3 Expression of Maximum Power Angle at Voltage Stability
Limit   *105*
4.4 Receiving-end Bus Voltage in a Weak Transmission Line
under Normal Operating Condition   *106*
4.5 Expression of Limiting Reactive Power Requirement   *106*
4.6 Relation between Reactive Power Variation and System
Stability   *107*
4.7 Reactive Power Sensitivity Governed Voltage Stability   *109*
4.8 Voltage Stability and Load Flow Solution   *110*
4.9 Loading Limit of a Transmission System at Voltage
Stability   *112*

Exercises   *114*

5. VOLTAGE STABILITY INDICATORS   115

5.1 Introduction   *115*
5.2 A Fundamental Indicator Using $P-V$ and $Q-V$ Curves   *115*
5.3 Criteria of Voltage Stability   *117*

    5.3.1 dE/dV Criterion   *117*
    5.3.2 dQ/dV Criterion   *121*

5.4 Load Voltage Indicator for Voltage Stability   *123*
5.5 A Direct Indicator of Voltage Stability and its Implication on
Voltage Stability Margin   *125*
5.6 A Voltage Stability Indicator Based on Governing Effect of
Load Increase   *128*
5.7 Voltage Stability Index (L)   *129*
5.8 Singular Value Decomposition   *130*
5.9 Expressions for Different Indicators to Investigate the
Voltage Security of a Power System   *131*
5.10 Voltage Stability Evaluation by Using Maximum Power
Transfer Phasor Diagram   *133*
5.11 A Tool of Assessing Voltage Stability Limit   *136*
5.12 Effect of System Reactance and Load Power Factor on
Voltage Stability   *138*
5.13 Voltage Stability and its Relation with Off-Nominal Tap Ratio of
Load Side Transformer and Source-to-Load Reactances   *143*

Exercises   *147*

vi Contents

6. ASSESSMENT OF VOLTAGE STABILITY AND SECURITY   **148**

   6.1   Introduction   *148*
   6.2   Power System Security Analysis   *149*

        6.2.1   Analysis and Control of Operating States   *149*
        6.2.2   Planning and Operation   *150*
        6.2.3   Security Assessment   *151*

   6.3   Computation of Voltage Stability Limits   *154*

        6.3.1   Concept of Transfer Capacity   *154*
        6.3.2   Voltage Stability Margin   *155*
        6.3.3   Computation of Voltage Collapse Point   *156*
        6.3.4   Method of Minimum Singular Value   *156*
        6.3.5   Point of Collapse Method   *157*
        6.3.6   Optimisation Method   *158*
        6.3.7   Continuation Load-flow Method   *159*
        6.3.8   Computation of Maximum Loading Point   *160*
        6.3.9   Eigenvalue Analysis   *161*
        6.3.10   Bus Equivalencing Technique   *164*
        6.3.11   Method of Optimal Power Flow   *167*
        6.3.12   Method Using Stability Boundary in $P-Q$ Plane   *174*

   6.4   Contingency Analysis   *176*

        6.4.1   Determination of Power System Post-disturbance State   *176*
        6.4.2   Contingency Selection   *176*
        6.4.3   Contingency Ranking Methods for Voltage Stability Studies   *179*
        6.4.4   Comparison of Contingency Ranking Methods   *181*

   Exercises   *182*

7. VOLTAGE CONTROL AND IMPROVEMENT OF VOLTAGE STABILITY IN POWER TRANSMISSION SYSTEMS   **183**

   7.1   Introduction   *183*
   7.2   Role of Transformers in Voltage Control of a Power System   *184*

        7.2.1   Modelling of Transformers   *184*
        7.2.2   Case I: Voltage Control of a Radial Load   *185*
        7.2.3   Case II: Tie-Transformer between Systems of Various Strengths   *186*

   7.3   Quantitative Method to Determine the Tap Setting for Voltage Control Using OLTC at a Load Bus   *189*

        7.3.1   Reactive Power at Receiving Bus   *189*
        7.3.2   Method of Voltage Control by Tap-Changing Transformers   *189*

   7.4   Effect of On-Load Tap-changer Transformer on Voltage Stability   *192*
   7.5   Practical Aspects of Voltage Instability due to OLTC Operation   *195*
   7.6   Methods of Improving Voltage Stability   *197*

Contents **vii**

- 7.7 Enhancement of Localised Reactive Power Support 197
    - 7.7.1 Fixed Shunt Capacitor Compensation at the Receiving-end *197*
- 7.8 Series Compensation *202*
    - 7.8.1 Concepts in Loadability and Voltage Stability of Series Compensated Lines *203*
    - 7.8.2 Effect of Series Compensation on the Voltage Instability of EHV Long Lines *210*
    - 7.8.3 Relation between the Receiving-end Voltage and the Degree of Compensation *211*
    - 7.8.4 Relation between E and V for Voltage Stability *212*
    - 7.8.5 Effect of Location of Series Capacitor on Voltage Stability *214*
    - 7.8.6 Effect of Capacitor Bussing Arrangements and Contingency on Voltage Stability of Series Compensated Lines *219*
- 7.9 An Optimal Load Shedding Scheme to Maintain Voltage Stability Following a Line Contingency *223*
    - 7.9.1 Introduction *223*
    - 7.9.2 Analysis *224*
- 7.10 Synchronous Condenser at the Load Bus *228*
- 7.11 FACTS Devices *232*
- 7.12 Classification of FACTS Controllers *233*
- 7.13 Series Controllers *234*
- 7.14 Type of Series Connected Controllers *234*
    - 7.14.1 SSSC *234*
    - 7.14.2 TSSC and TCSC Combined Controller *236*
    - 7.14.3 Power Flow Model of TCSC *236*
    - 7.14.4 TCSR *237*
- 7.15 Shunt Controllers *238*
- 7.16 Types of Shunt Controllers *238*
    - 7.16.1 SVC *238*
    - 7.16.2 SSG *239*
    - 7.16.3 Power Flow Model of SVC *239*
    - 7.16.4 STATCOM *241*
    - 7.16.5 Power Flow Model of STATCOM *241*
- 7.17 Series–Series Controllers *243*
- 7.18 Combined Shunt–Series Connected Controllers *243*
    - 7.18.1 Static Phase Shifter (SPS) *243*
    - 7.18.2 Unified Power Flow Controller (UPFC) *246*
    - 7.18.3 UPFC Modelling *248*
    - 7.18.4 Interphase Power Controller (IPC) *250*
- 7.19 Advantages of FACTS Devices *251*
- *Exercises* *251*

References 253

Index *257*

Preface

The rapid growth of electrical power transmission since the Second World War has ushered in the problem of stability of power systems so much so that the topic of power system stability has today become an important sub-branch of electrical engineering. The pioneering work of authors like E.W. Kimbark, S.B. Crary, Edith Clark, etc. marked the first milestones on the way to extremely rapid development of this sub-branch. During the last 30 years, several papers and conference reports on power system stability have been published throughout the world, indicating the continuing interest of electrical engineers in this field. During these years, this sub-branch of electrical engineering has also come in quite close contact with a number of related disciplines such as 'network theory', 'optimal control theory', 'digital techniques', 'electrical machines', etc.

The number of books dealing only with the problem of power system stability is, however, rather few, especially in consideration of the rapid growth of this topic in recent years. Most of the advanced books devote only a chapter to the topic of power system stability. However, the versatile concepts used in different aspects of stability need a much more thorough discussion. This fact stimulated the authors to make an effort to prepare a book that would cover fairly the concepts and problems of power system stability.

Thirty years ago, the presence of reactive power in the power system was considered a relatively trouble free phenomenon. The control of voltage could be done in a de-centralized manner by regulators, keeping the pre-assigned set points of voltage levels at certain load buses using the local reactive sources such as synchronous condenser and shunt capacitors. Things started to change around 25 years ago when the real importance of the problem was felt due to some reactive stability problem becoming visible in heavily-loaded long transmission systems. Reactive power problems lead to the phenomenon of voltage instability. Though this problem is not new to the power system engineers and researchers, several researches have been conducted in the last 20 years and extensive contributions have been made towards understanding, analyzing, predicting, compensating and simulating the voltage stability perspective of reactive power, and hence voltage stability has become an important factor in every major transmission network planning and operation.

The voltage instability phenomenon was well recognized in radial distribution systems and has been explained in many textbooks. It was not, however, a problem that the transmission planners had to deal with extensively until recently. Most of the early developments of the major HV and EHV networks faced the classical machine angle stability problem. Innovation in both analytical techniques and stabilizing measures made it possible to maximize the power transfer capabilities of the transmission systems. The result was increasingly transfer of power over long distances of transmission.

Reactive power allocation studies (sometimes referred to as VAR allocation studies) have been routinely done to determine the reactive support required to meet the future demand while maintaining an acceptable voltage profile under normal conditions and credible contingencies. Proximity to voltage stability limits has been recognized recently by many utilities. Some experienced actual collapse while others observed the symptoms during planning studies; thus voltage stability has become an important factor in planning and operation of major transmission networks.

While the basic phenomenon as applied to a radial feeder is well understood and formulated, it is not as simple in large networks. The large networks present major challenges in establishing a sound and simple analytical procedure, and in formulating a quantitative measure of proximity to voltage instability and adequate operating margins. It is recognized that voltage stability and voltage collapse are associated with relatively slow variation in load, and network and control characteristics. As will be seen later in this book, the slow nature of this variation has created a debate in the literature on whether a steady-state approach would be adequate for the analysis. This is not to say that the phenomenon is static but the issue on the other hand is to find whether it is possible to account for dynamic factors in steady-state base indicators. It should also be emphasized that there is no lack of detailed dynamic models for all the relevant elements affecting voltage stability—long-term stability programs are capable of modelling slope controllers such as on-load tap changers (OLTC), automatic generation control (AGC), under-frequency and undervoltage load shedding, generator field current limiters, etc. This approach, however, is time consuming for mass production and does not readily provide a simple quantitative index.

In writing this book the authors have assumed that the readers have the basic knowledge of power systems including the ABC of power system stability. In this book the basic concepts of power system stability and other operational aspects have been reviewed in brief and an effort has been made to highlight the advanced as well as the practical aspects of stability which have so far been ignored in the conventional textbooks. Modern concepts and applications have been introduced; theoretical as well as stimulated studies have been presented whenever necessary. Special attention has been given to reactive power control and stability aspects.

In this book the most difficult task of presenting a complete overview of the research and engineering aspects of the problem of stability, suitable both for academics and practising engineers, has been attempted along with a brief historical review of the concerned topics. In some instances the authors have included some of their own research results while maintaining the uniformity of the overall treatment of the problem. It is expected that the book will be beneficial to academics, research scholars, postgraduate students and practising engineers.

The authors acknowledge the contribution of various research scholars, particularly the assistance provided by Ms. Sawan Sen, Ms. Priyanka Roy, and Ms. Poulomi Mitra in shaping the manuscript. The authors also acknowledge the patience and understanding of their spouses and for rendering all emotional encouragement while the manuscript was being prepared. The effort of the staff of PHI Learning towards speedy processing of the manuscript is highly appreciated by the authors.

Abhijit Chakrabarti
D.P. Kothari
A.K. Mukhopadhyay
Abhinandan De

Chapter 1

Power Transmission in Uncompensated AC Transmission Lines

1.1 INTRODUCTION

It is very important to understand the implications of reactive power in voltage control of transmission systems. The reactive power is essential for the operation of electromagnetic energy devices. It constitutes the voltage and current loading of circuits but does not result in active power consumption when the devices are either purely inductive or purely capacitive. Reactive power is the latent soul of operation of almost all of the electromagnetic energy devices. Electromagnetic devices draw lagging currents, thereby resulting in positive values of reactive power (Q). It may then be inferred that these devices 'absorb' reactive power. Electrostatic devices, on the other hand, store electric energy in fields. These devices draw leading currents and result in negative reactive power (Q). They 'supply' reactive power. In this chapter, an understanding of reactive power associated with power transmission systems is presented and for the sake of clear understanding and ease of calculations, presently we assume lossless lines.

1.2 ELECTRICAL PARAMETERS OF TRANSMISSION LINES

Series resistance (r ohm/km), series inductance (l henry/km), shunt conductance (g mho/km) and capacitance (c farad/km) are the four basic parameters (commonly known as *primary constants*) of any transmission line. Representing the net series impedance and shunt admittance by z and y for the unit length of the line respectively, we have

$$z = (r + j\omega l) \text{ ohm/km} \qquad (1.1)$$
$$y = (g + j\omega c) \text{ mho/km} \qquad (1.2)$$

Z and Y represent the total series impedance and shunt admittance and are obtained by multiplying z and y with line length.

All the four electrical parameters are the functions of the overhead conductor size, type, spacing, height above ground, frequency and thermal rating. Often, two three-phase power lines are electrically parallel. The line parameters are affected by the mutual action of one upon the other. Use of bundle conductors and unsymmetrical conductor spacing have also been found to affect the reactance parameter of the line.

It is a reasonable approximation if the shunt conductance and the series resistance are neglected as the behaviour of a line is dominated by the series inductance and shunt capacitance. It is intended here to discuss their role in the context of power flow and voltage stability.

1.3 REPRESENTATION OF A TRANSMISSION LINE AND TRANSMISSION LINE EQUATION

Transmission lines are generally classified as short, medium and long lines as their representations are dependent on their length. Though in an actual line the line parameters are evenly distributed (Figure 1.1), short and medium transmission lines can be better analysed utilising the lumped parameter model.

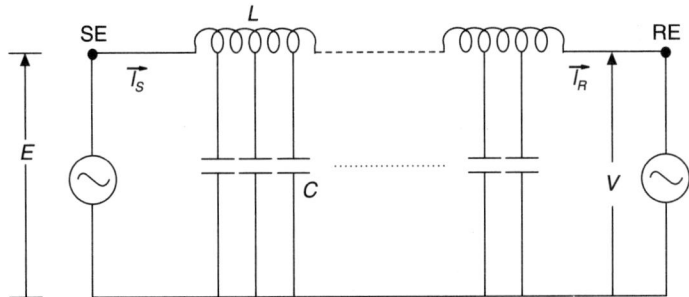

Figure 1.1 Circuit representation of transmission line parameters.

The short line can be represented by the simple positive sequence equivalent circuit shown in Figure 1.2. It may be assumed as a four-terminal two-port network where it can be shown that

$$A = 1, B = Z, C = 0, D = 1$$

For medium lines it is customary to take the distributed capacitance of the line in the form of a lumped shunt capacitance, as shown in the equivalent circuit of a medium line (Figure 1.3). Any one of the two standard versions of representation (i.e. π or T) may be adopted as there is little difference between the two from the point of view of accuracy.

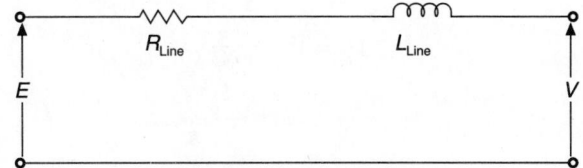

Figure 1.2 Positive sequence equivalent circuit of a short line.

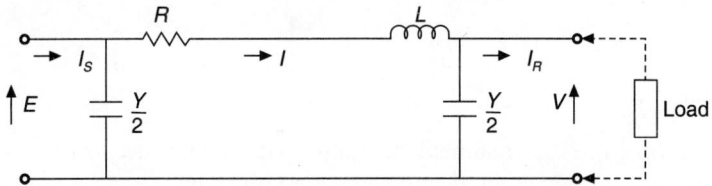

Figure 1.3(a) Lumped parameter π equivalent circuit of a medium line.

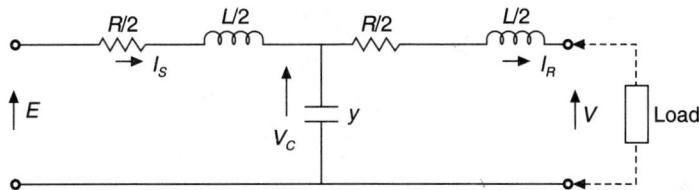

Figure 1.3(b) Lumped parameter T equivalent circuit of a medium line.

For the π-network representation,

$$E = V + IZ \tag{1.3}$$

$$I_S = I + E\frac{Y}{2} \tag{1.4}$$

$$I = I_R + V\frac{Y}{2} \tag{1.5}$$

and
$$A = D = 1 + \frac{YZ}{2} \tag{1.6}$$

$$B = Z \tag{1.7}$$

$$C = \left(1 + \frac{ZY}{4}\right)Y \tag{1.8}$$

Similarly for the T-network,

$$E = V_C + \frac{ZI_S}{2} \tag{1.9}$$

$$V_C = V + \frac{ZI_R}{2} \qquad (1.10)$$

$$I_S = I_R + V_C Y \qquad (1.11)$$

and

$$A = D = 1 + \frac{ZY}{2} \qquad (1.12)$$

$$B = \left(1 + \frac{ZY}{2}\right) Z \qquad (1.13)$$

$$C = Y \qquad (1.14)$$

For the long line, the treatment incorporates the *distributed parameters*. Here, the voltage and current at the receiving end are given by

$$V = E \cosh \gamma L - I_S Z_0 \sinh \gamma L \qquad (1.15)$$

and

$$I_R = I_S \cosh \gamma L - \frac{E}{Z_0} \sinh \gamma L \qquad (1.16)$$

where γ, the *propagation constant* is given by

$$\gamma = \sqrt{zy} = \sqrt{(r + j\omega l)(g + j\omega c)} \qquad (1.17)$$

while Z_0, the characteristic impedance is given by

$$\sqrt{\frac{z}{y}} = \sqrt{\frac{r + j\omega l}{g + j\omega c}} \qquad (1.18)$$

In a lossless line*, α, the *attenuation constant* is zero; the propagation constant can then be represented as

$$\gamma (= \alpha + j\beta) = j\beta \qquad (\beta \text{ being the } phase\ constant) \qquad (1.19)$$

Also,

$$\beta = \frac{1}{2\pi\sqrt{lc}} \qquad (1.20)$$

The sending end voltage and current in a long line may be expressed as

$$E = V \cosh \gamma L + I_R Z_0 \sinh \gamma L \qquad (1.21)$$

and

$$I_S = \frac{V}{Z_0} \sinh \gamma L + I_R \cosh \gamma L \qquad (1.22)$$

(L being the length of the line).

---

*For lossless line, $R = 0$; $g = 0$ since the line is assumed to have no series resistance and zero shunt conductance.

The parameters of the equivalent four-terminal network are, therefore,

$$A = D = \cosh\sqrt{YZ} \tag{1.23}$$

$$B = \sqrt{\frac{Z}{Y}}\sinh\sqrt{ZY} \tag{1.24}$$

$$C = \sqrt{\frac{Y}{Z}}\sinh\sqrt{YZ} \tag{1.25}$$

Utilising the standard versions of hyperbolic functions, Eqs. (1.23), (1.24) and (1.25) reduce to,

$$\Rightarrow \quad A = D \approx \left(1 + \frac{ZY}{2}\right) \tag{1.26}$$

$$\Rightarrow \quad B \approx Z\left(1 + \frac{YZ}{6}\right) \tag{1.27}$$

$$\Rightarrow \quad C \approx Y\left(1 + \frac{YZ}{6}\right) \tag{1.28}$$

Also, for the lossless line, the expression of voltage and current at any point x (distance from the sending end) is given by

$$V_x = V\cos\beta(L-x) + jZ_0 I_R \sin\beta(L-x) \tag{1.29}$$

and

$$I_x = j\left[\frac{V}{Z_0}\right]\sin\beta(L-x) + I_R \cos\beta(L-x) \tag{1.30}$$

The quantity βx represents the *electrical length* of the line and is expressed in radian (symbol being θ).

1.4 CONCEPT OF POWER IN AC TRANSMISSION SYSTEMS

Conventionally, the total power flow in a single-phase AC system is termed 'Complex power' or 'Apparent power' and is represented by kVA (or MVA) and consequently, the rating of any power apparatus is expressed in terms of kVA (or MVA).

For a single-phase AC circuit, the voltage and current phasors are conventionally written as

$$V = |V|e^{j\angle V} \tag{1.31}$$

and

$$I = |I|e^{j\angle I} \tag{1.32}$$

Introducing the complex conjugate of current as

$$I^* = |I|e^{-j\angle I} \tag{1.32a}$$

the complex power S is defined by

$$S = VI^* \tag{1.33}$$

Substituting Eqs. (1.31) and (1.32a) in (1.33), we get

$$S = VI^* = |V||I| e^{j(\angle V - \angle I)}$$

$$= |V||I| e^{j\phi} \tag{1.34}$$

where the phase angle $\phi = (\angle V - \angle I)$ represents the angle between the voltage and current.

Thus Eq. (1.34) becomes

$$S = |V||I| \cos\phi + j|V||I| \sin\phi$$

i.e.
$$S = (P + jQ) \tag{1.35}$$

where $P = |V||I| \cos\phi$ and $Q = |V||I| \sin\phi$.

P and Q represent the *real* and *imaginary* parts of the complex power S, and are popularly known as 'real power' and 'reactive power' respectively.

From Eq. (1.33), alternative expressions for complex power may be obtained using the basic relationship between V and I as

$$\left.\begin{array}{l} V = ZI \\ I = YV \end{array}\right\} \tag{1.36}$$

Thus,
$$\left.\begin{array}{l} S = VI^* = VY^* V^* = Y^* |V|^2 \\ = IZI^* = Z|I|^2 \end{array}\right\} \tag{1.37}$$

In power system analysis, expressions of complex power S may be expressed in any one of the forms represented by Eq. (1.35) or Eq. (1.37).

It is also evident that

$$|S| = |VI^*| = |V||I| = \sqrt{P^2 + Q^2} \tag{1.38}$$

P and Q both have the units 'watt' (or kilowatt) but to emphasize the fact that Q is associated with the reactive or wattless component of current, the unit of Q is frequently termed voltamp reactive (or VAR). S being a complex quantity, it is related to P and Q and has been diagrammatically shown in Figure 1.4 (utilising Eq. (1.38) as a power triangle.

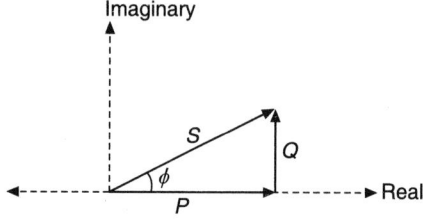

Figure 1.4 Power triangle.

1.5 POWER FLOW IN A TWO-TERMINAL POWER TRANSMISSION NETWORK

Figure 1.5 represents the equivalent T-section of a two-terminal power transmission system where Z_1 and Z_2 represent the series arm and Z_3, the shunt arm of the T-section. The real and reactive power at both the ends are then given by

$$P_1 = \text{real part of } V_1 I_1^* = V_1 I_1 \cos\phi_1 \tag{1.39}$$

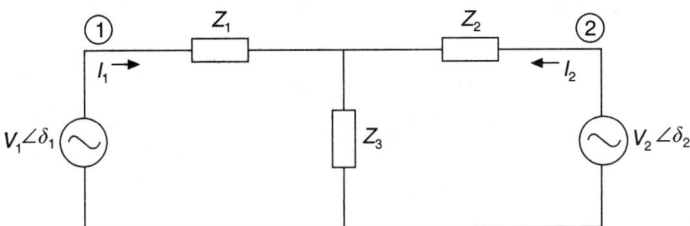

Figure 1.5 Equivalent T-section of a two-terminal power transmission system.

$$Q_1 = \text{imaginary part of } V_1 I_1^* = V_1 I_1 \sin\phi_1 \tag{1.40}$$

and

$$P_2 = \text{real part of } V_2 I_2^* = V_2 I_2 \cos\phi_2 \tag{1.41}$$

$$Q_2 = \text{imaginary part of } V_2 I_2^* = V_2 I_2 \sin\phi_2 \tag{1.42}$$

where, P and Q represent the real and reactive powers at the two ends; V and I, the respective voltages and currents with their phase angle ϕ_1 and ϕ_2. Suffixes 1 and 2 have been added to denote the two ends of the power system.

If Y_{11} (= $Y_1 \angle -\theta_{11}$) and Y_{12} (= $Y_2 \angle -\theta_{12}$) represent the short-circuit driving point admittance and transfer admittance†, the expressions for the line currents at the two ends are given by

$$I_1 = V_1 Y_{11} - V_2 Y_{12} \tag{1.43}$$
$$I_2 = V_2 Y_{22} - V_1 Y_{21} \tag{1.44}$$

Assuming the phasors V_1 and V_2 to be operating at angles δ_1 and δ_2, the real power injecting at bus (1) can be obtained by utilising Eqs. (1.39) and (1.43) as

$$P_1 = (V_1 e^{j\delta_1})(V_1 e^{j\delta_1} Y_{11} e^{-j\theta_{11}} - V_2 e^{j\delta_2} Y_{12} e^{-j\theta_{12}})^*$$
$$= V_1 e^{j\delta_1} [V_1 Y_{11}\{\cos(\delta_1 - \theta_{11}) + j\sin(\delta_1 - \theta_{11})\}$$
$$- V_2 Y_{12}\{\cos(\delta_2 - \theta_{12}) + j\sin(\delta_2 - \theta_{12})\}]^* \tag{1.45}$$

Again, assuming $\delta_{12} = \delta_1 - \delta_2$, $\alpha_{11} = (90 - \theta_{11})$ and $\alpha_{12} = (90 - \theta_{12})$, trigonometric simplification of Eq. (1.45) yields

---

†For convenience the mod signs denoting magnitude are removed.

$$P_1 = V_1^2 Y_{11} \sin\alpha_{11} + V_1 V_2 Y_{12} \sin(\delta_{12} - \alpha_{12}) \tag{1.46}$$

Similarly, the expression for the real power injection at bus (2) can be obtained by utilising Eqs. (1.41) and (1.44) and assuming $\delta_{12} = -\delta_{21} = (\delta_1 - \delta_2)$, $\alpha_{22} = (90° - \theta_{22})$, $\alpha_{12} = \alpha_{21} = (90° - \theta_{12}) = (90° - \theta_{21})$. We then have

$$P_2 = -V_2^2 Y_{22} \sin\alpha_{22} - V_1 V_2 Y_{21} \sin(\delta_{12} + \alpha_{12}) \tag{1.47}$$

Again, utilising the expression of I_1 from Eq. (1.43) in Eq. (1.40), simplification yields the value of Q_1 as

$$Q_1 = V_1^2 Y_{11} \cos\alpha_{11} - V_1 V_2 Y_{12} \cos(\delta_{12} - \alpha_{12}) \tag{1.48}$$

while from Eqs. (1.44) and (1.42), the value of Q_2 has been obtained as

$$Q_2 = -V_2^2 Y_{22} \cos\alpha_{22} + V_1 V_2 Y_{12} \cos(\delta_{12} + \alpha_{21}) \tag{1.49}$$

In EHV lines the (r/x) ratio of conductors is small, and in analysis it is a reasonable assumption to neglect the resistance in the line. In such a case when there is no resistance in the equivalent network connecting the two buses,

$$\alpha_{11} = 0, \ \alpha_{12} = 0, \ \alpha_{22} = 0, \ \alpha_{21} = 0$$

Thus, the modified expressions (in the lossless frame) for real and reactive powers at the buses (1) and (2) are given by

$$\left.\begin{aligned} P_1 &= V_1 V_2 Y_{12} \sin\delta_{12} \\ P_2 &= -V_1 V_2 Y_{21} \sin\delta_{12} \\ Q_1 &= -V_1 V_2 Y_{12} \cos\delta_{12} + V_1^2 Y_{11} \\ Q_2 &= V_1 V_2 Y_{12} \cos\delta_{12} - V_2^2 Y_{22} \end{aligned}\right\} \tag{1.50}$$

EXAMPLE 1.1 Figure E1.1 represents the π-equivalent of a power line with the series and shunt admittance branches. Find the values of P_1, P_2, Q_1, and Q_2 at different values of the power angle.

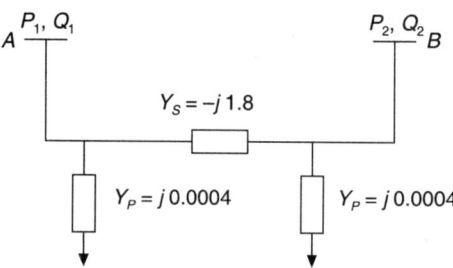

Figure E1.1

Solution Here in Figure E1.1,

$$Y_{AA} \angle\alpha_2 = Y_S + Y_P = -j1.8 + j0.0004 = -j1.7996$$
$$= 1.7996 \angle -90°$$

$$Y_{AB} \angle \alpha_1 = -Y_S = j1.8 = 1.8 \angle 90°$$
$$\alpha_{AB} = 90° - \alpha_1 = 90° - 90° = 0°$$
$$\alpha_{AA} = 90° - \alpha_2 = 90° - (-90°) = 90° + 90° = 180°$$

Since the circuit is symmetrical,

$$Y_{BB} = Y_{AA} \text{ and } Y_{AB} = Y_{BA}$$
$$\alpha_{BB} = \alpha_{AA} \text{ and } \alpha_{AB} = \alpha_{BA}$$
$$Y_{AA} = Y_{BB} = 1.7996, E_A = E_B = 1.00$$
$$Y_{AB} = Y_{BA} = 1.8$$
$$\alpha_{AA} = \alpha_{BB} = 180°$$
$$\alpha_{AB} = \alpha_{BA} = 0°$$

∴
$$P_1 = E_A^2 Y_{AA} \sin \alpha_{AA} + E_A E_B Y_{AB} \sin(\delta_{AB} - \alpha_{AB})$$
$$= 1.7996 \sin 180° + 1.8 \sin(\delta_{AB} - 0°)$$
$$= 1.8 \sin \delta_{AB}$$

∴
$$P_2 = -E_B^2 Y_{BB} \sin \alpha_{BB} - E_A E_B Y_{AB} \sin(\delta_{AB} + \alpha_{AB})$$
$$= -1.7996 \sin 180° - 1.8 \sin(\delta_{AB} + 0°)$$
$$= -1.8 \sin \delta_{AB}$$

∴
$$Q_1 = E_A^2 Y_{AA} \cos \alpha_{AA} + E_A E_B Y_{AB} \cos(\delta_{AB} - \alpha_{AB})$$
$$= 1.7996 \cos 180° - 1.8 \cos(\delta_{AB} - 0°)$$
$$= -1.7996 - 1.8 \cos \delta_{AB}$$

∴
$$Q_2 = -E_B^2 Y_{BB} \cos \alpha_{BB} + E_A E_B Y_{AB} \cos(\delta_{AB} + \alpha_{AB})$$
$$= -1.7996 \cos 180° + 1.8 \cos(\delta_{AB} + 0°)$$
$$= 1.7996 + 1.8 \cos \delta_{AB}$$

Table 1.1 gives the values of P_1, P_2, Q_1 and Q_2 for different power analysis of Example 1.1.

EXAMPLE 1.2 Repeat Example 1.1 for another equivalent π-section of a line shown in Figure E1.2.

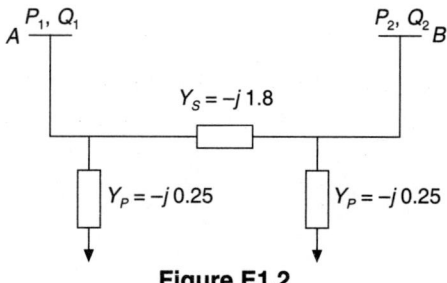

Figure E1.2

Table 1.1 Values of P_1, P_2, Q_1 and Q_2 for different values of power angle (Example 1.1)

δ_{AB}	$-P_2 = P_1$	Q_1	$Q_2 = -Q_1$
0	0.0000	−3.5996	3.5996
10	0.3125	−3.5722	3.5722
20	0.6156	−3.4910	3.4910
30	0.9000	−3.3584	3.3584
40	1.1570	−3.1784	3.1784
50	1.3788	−2.9566	2.9566
60	1.5588	−2.6996	2.6996
70	1.6914	−2.4152	2.4152
80	1.7726	−2.1121	2.1121
90	1.8000	−1.7996	1.7996
110	1.6914	−1.1839	1.1839
130	1.3788	−0.6425	0.6425
150	0.9000	−0.2407	0.2407
170	0.3125	−0.0269	0.0269
180	0.0000	−0.0004	0.0004
200	−0.6156	−0.1081	0.1081
220	−1.1570	−0.4207	0.4207
240	−1.5588	−0.8996	0.8996
260	−1.7726	−1.4870	1.4870
280	−1.7726	−2.1121	2.1121
300	−1.5588	−2.6996	2.6996
320	−1.1570	−3.1784	3.1784
340	−0.6156	−3.4910	3.4910
360	0.0000	−3.5996	3.5996

Solution

$$Y_{AA} \angle \alpha_2 = Y_S + Y_P = -j1.8 - j\,0.25$$
$$= -j2.05 = 2.05 \angle -90° \text{ p.u.}$$
$$Y_{AB} \angle \alpha_1 = -Y_S = j1.8 = 1.8 \angle 90° \text{ p.u.}$$
$$\alpha_{AB} = 90° - \alpha_1 = 90° - 90° = 0°$$
$$\alpha_{AA} = 90° - \alpha_2 = 90° - (-90°) = 180°$$

Since the circuit is symmetrical,

$$Y_{BB} = Y_{AA} \qquad Y_{AB} = Y_{BA}$$
$$\alpha_{BB} = \alpha_{AA} \qquad \alpha_{AB} = \alpha_{BA}$$
$$Y_{AA} = Y_{BB} = 2.05$$
$$Y_{AB} = Y_{BA} = 1.8$$
$$\alpha_{AA} = \alpha_{BB} = 180°$$
$$\alpha_{AB} = \alpha_{BA} = 0°$$

$$\therefore \quad P_1 = E_A^2 Y_{AA} \sin\alpha_{AA} + E_A E_B Y_{AB} \sin(\delta_{AB} - \alpha_{AB})$$
$$= 2.05 \sin 180° + 1.8 \sin\delta_{AB}$$
$$= 1.8 \sin\delta_{AB}$$

$$\therefore \quad P_2 = -E_B^2 Y_{BB} \sin\alpha_{BB} - E_A E_B Y_{AB} \sin(\delta_{AB} + \alpha_{AB})$$
$$= -2.05 \sin 180° - 1.8 \sin\delta_{AB}$$
$$= -1.8 \sin\delta_{AB}$$

$$\therefore \quad Q_1 = E_A^2 Y_{AA} \cos\alpha_{AA} - E_A E_B Y_{AB} \cos(\delta_{AB} - \alpha_{AB})$$
$$= 2.05 \cos 180° - 1.8 \cos\delta_{AB}$$
$$= -2.05 - 1.8 \cos\delta_{AB}$$

$$\therefore \quad Q_2 = -E_B^2 Y_{BB} \cos\alpha_{BB} + E_A E_B Y_{AB} \cos(\delta_{AB} + \alpha_{AB})$$
$$= -2.05 \cos 180° + 1.8 \cos\delta_{AB}$$
$$= 2.05 + 1.8 \cos\delta_{AB}$$

Table 1.2 gives the values of P_1, P_2, Q_1 and Q_2 for different power analysis of Example 1.2.

1.6 RECEIVING-END POWER CIRCLE DIAGRAM

Assuming the sending-end voltage to be V_S and the receiving-end voltage to be V_R, in terms of transmission parameters we can express the sending-end voltage as

$$V_S = AV_R + BI_R$$

or

$$I_R = \frac{V_S - AV_R}{B} \qquad (1.51)$$

Since
$$A = |A|\angle\alpha; \quad B = |B|\angle\beta,$$
$$V_R = |V_R|\angle 0° \text{ (the reference phasor), and}$$
$$V_S = |V_S|\angle\delta,$$

we can write:

$$I_R = \frac{|V_S|}{|B|}\angle\delta - \beta - \frac{|A||V_R|}{|B|}\angle\alpha - \beta$$

$$\therefore \quad S_R (= P_R + jQ_R) = V_R I_R^*$$
$$= \frac{|V_S||V_R|}{|B|}\angle\beta - \delta - \frac{|A||V_R|^2}{|B|}\angle\beta - \alpha \qquad (1.52)$$

Table 1.2 Values of P_1, P_2, Q_1 and Q_2 for different values of power angle (Example 1.2)

δ_{AB}	P_1	$P_2(-P_1)$	Q_1	$Q_2 = -Q_1$
0	0.0000	↑	−3.8500	3.8500
10	0.3125		−3.8226	3.8226
20	0.6156		−3.7414	3.7414
30	0.9000		−3.6088	3.6088
40	1.1570		−3.4288	3.4288
50	1.3788		−3.2070	3.2070
60	1.5588		−2.9500	2.9500
70	1.6914		−2.6652	2.6652
80	1.7726		−2.3625	2.3625
90	1.8000		−2.0500	2.0500
110	1.6914		−1.4343	1.4343
130	1.3788		−0.8929	0.8929
150	0.9000		−0.4919	0.4919
170	0.3125	Same as	−0.2773	0.2773
180	0.0000	P_1	−0.2500	0.2500
200	−0.6156		−0.3585	0.3585
220	−1.1570		−0.6711	0.6711
240	−1.5588		−1.1500	1.1500
260	−1.7726		−1.7374	1.7374
280	−1.7726		−2.3625	2.3625
300	−1.5588		−2.9500	2.9500
320	−1.1570		−3.4288	3.4288
340	−0.6156		−3.7414	3.7414
360	0.0000	↓	−3.8500	3.8500

The real and reactive powers at the receiving end, thus, can be expressed as

$$\left.\begin{aligned} P_R &= \frac{|V_S||V_R|}{|B|}\cos(\beta - \delta) - \frac{|A||V_R|^2}{|B|}\cos(\beta - \alpha) \\ Q_R &= \frac{|V_S||V_R|}{|B|}\sin(\beta - \delta) - \frac{|A||V_R|^2}{|B|}\sin(\beta - \alpha) \end{aligned}\right\} \quad (1.53)$$

We thus see that S_R becomes a resultant of combination of two phasors (Eq. (1.52)). We can plot these phasors in the complex phase where the horizontal and vertical co-ordinates are "Watts" and "VARs" respectively. Figure 1.6 exhibits the two complex quantities and their difference.

The same phasors have been plotted in Figure 1.7 with the origin of the co-ordinate axes shifted. This is a power diagram with plotting of the respective phasors with their resultant $|P_R + jQ_R|$ replaced by $|V_R||I_R|$ (ϕ_R being the angle of the resultant with the horizontal axis). Thus we can write for this representation,

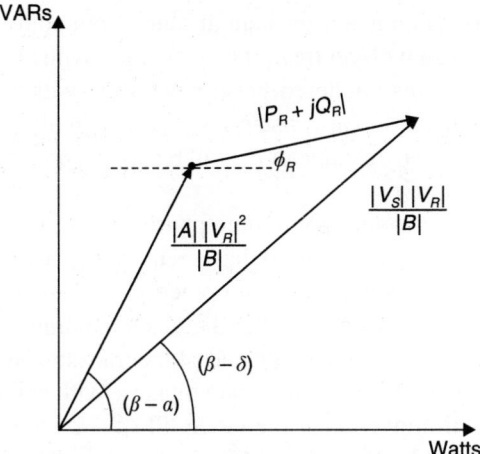

Figure 1.6 Phasors plotted in complex plane.

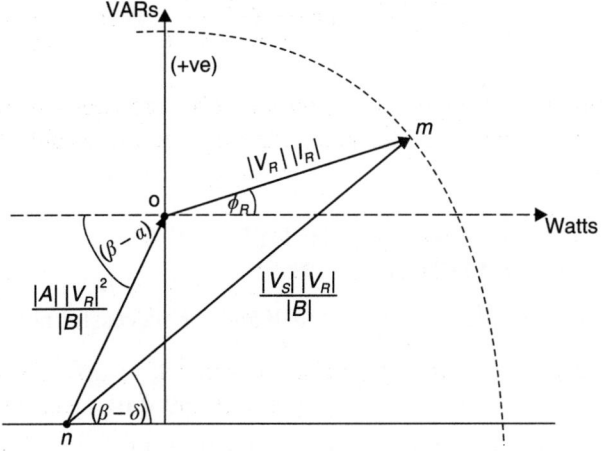

Figure 1.7 Receiving end power circle diagram obtained by shifting the origin of the co-ordinate axes.

$$P_R = |V_R||I_R|\cos\phi_R$$

$$Q_R = |V_R||I_R|\sin\phi_R$$

[Q is +ve with I lagging V]

We now can note the following things:
(i) The position of the point n is independent of the receiving end current I_R.
(ii) The position of n will change if $|V_R|$ is altered.
(iii) The distance nm in Figure 1.7 is constant for fixed values of $|V_S|$ and $|V_R|$.
(iv) With the change in load, the distance om would change (change in location of m indicates variation of the connected load at the receiving end).

With point m remaining at a constant distance from point n, V is constrained to follow a circular path with centre n. If $|V_S|$ changes with $|V_R|$ remaining same, the location of n remains unaltered but a new circle with a new radius nm is formed. Thus, change in $|V_S|$ at constant $|V_R|$ results in different parallel circles with centres at n. Also, any change in P_R will result in a corresponding change in Q_R to keep m on the circle.

When making calculations for a three-phase system, it is advisable to work everything on a single-phase basis. All that is required to be done is to convert the the line voltage to phase voltage and then proceed.

From Figure 1.7 it may be seen that there is a maximum limit to the power that can be transmitted to the receiving end of the transmission line for specified magnitudes of sending- and receiving-end voltages. With enhancement in power delivery, point m will move along the circle till $(\beta - \delta)$ approaches zero. This means that for $\beta - \delta = 0$, i.e. for $\delta = \beta$, maximum power is delivered. The maximum power is obtained on

$$P_{R(max)} = \frac{|V_S||V_R|}{|B|} - \frac{|A||V_R|^2}{|B|} \cos(\beta - \alpha) \tag{1.54}$$

However, to attain this condition, the load must operate at leading power factor and hence operation is usually limited by keeping δ at a low value (say up to 20° or 30°); also $|V_R|$ is kept equal to $|V_S|$ or very close to $|V_S|$.

1.7 REACTIVE POWER REQUIREMENT OF AN UNCOMPENSATED LINE

Let δ be the phase angle between E and V and for successful power transfer, E leads V by δ.

$$E = |E|e^{j\delta} = E(\cos\delta + j\sin\delta) \tag{1.55}$$

In the lossless frame, with $\gamma = j\beta$, Eq. (1.21) reduces to

$$E = V\cos\beta L + I_R Z_0 \sin\beta L$$

i.e.
$$E = V\cos\theta + I_R Z_0 \sin\theta \tag{1.56}$$

Comparison of Eq. (1.55) with (1.56) yields

$$E(\cos\delta + j\sin\delta) = V\cos\theta + jZ_0 \sin\theta \cdot \frac{P - jQ}{V} \tag{1.57}$$

where
$$I_R = \frac{P - jQ}{V}$$

Equating the real and imaginary parts, Eq. (1.57) becomes

$$E\cos\delta = V\cos\theta + Z_0 \frac{Q}{V}\sin\theta \tag{1.58}$$

and
$$E\sin\delta = Z_0 \frac{P}{V}\sin\theta \tag{1.59}$$

Rearrangement of Eq. (1.58) yields the expression of receiving-end reactive power as

$$Q_R = \frac{EV \cos\delta}{Z_0 \sin\theta} - \frac{V^2 \cos\theta}{Z_0 \sin\theta} \quad (1.60)$$

Similarly, the sending-end reactive power is obtained as

$$Q_S = -\left[\frac{EV \cos\delta}{Z_0 \sin\theta} - \frac{E^2 \cos\theta}{Z_0 \sin\theta}\right] \quad (1.61)$$

Expressions (1.60) and (1.61) provide the terminal reactive power when the terminal voltages are E and V at the sending and receiving ends respectively.

However when $E = V$, the expression becomes

$$Q_R = \frac{E^2 \cos\delta - E^2 \cos\theta}{Z_0 \sin\theta} \quad (1.62)$$

and

$$Q_S = -\left[\frac{E^2 \cos\delta - E^2 \cos\theta}{Z_0 \sin\theta}\right] \quad (1.63)$$

Equations (1.62) and (1.63) reveal that $Q_R = -Q_S$.

For short lossless lines, θ is very small and hence $\cos\theta \simeq 1$ while $Z_0 \sin\theta \simeq X$ (line). Hence Eq. (1.63) reduces to

$$Q_S = \frac{E^2 (1-\cos\delta)}{X} = -Q_R \quad \text{assuming } E = V \quad (1.64)$$

For a small amount of real power transfer, the reactive power is surplus in the line and hence the terminal devices need to absorb the reactive power. This can be obtained by simulating Eqs. (1.62) and (1.63) with low values of δ and keeping $E = 1.00$ p.u. The generation of reactive power is more for longer lines.

On the other hand, if the power flow is heavy, the line's reactive power demand increases and terminal devices are needed to supply reactive power to the line. This can be obtained by simulating Eqs. (1.62) and (1.63) with higher values of δ and keeping $E = 1.00$. More reactive power needs to be supplied for longer line lengths and when they are heavily loaded.

The reactive power profile of the line for a particular line loading is plotted in Figure 1.8 which shows that the characteristic crosses the X-axis at a particular point. This point represents the magnitude of real power at which the reactive power requirement is nil at the terminals. This means that under this condition, the reactive power generated by the line is absorbed by itself and hence this particular point on the real power axis is said to be the surge impedance loading (SIL) point.

It is also evident from Figure 1.8 that for any particular amount of power transfer, there is also a specific line length that enables the line not to be constrained by the reactive power condition, i.e. the line can run without any external reactive

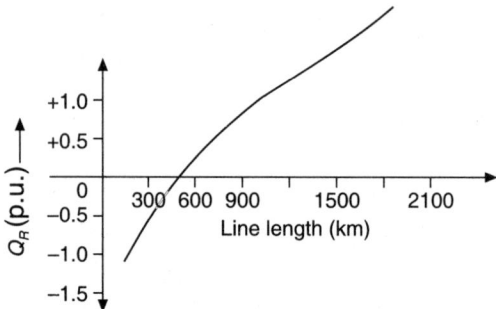

Figure 1.8 Reactive power profile of transmission line against the line length.

power support if the line can be operated with its real power value at a specific level (obviously this is the SIL).

1.8 IMPLICATION OF SURGE IMPEDANCE LOADING

The *surge impedance* is the apparent impedance of a long line at any point along the line. Hence,

$$Z_x = \frac{V_x}{I_x}$$

From Eqs. (1.29) and (1.30), we have

$$Z_x = \frac{V \cos \beta (L - x) + jZ_0 I_R \sin \beta (L - x)}{j (V/Z_0) \sin \beta (L - x) + I_R \cos \beta (L - x)}$$

$$= \frac{V [\cos \beta (L - x) + j \sin \beta (L - x)]}{V [\cos \beta (L - x) + j \sin \beta (L - x)]} Z_0 = Z_0 \quad (1.65)$$

Hence, it may be observed that the surge impedance of the line does not depend on the length of the line. The *characteristic impedance* Z_0 is known to be the *surge impedance* or *natural impedance* of a distortionless line.

Hence, $\qquad Z_0 \text{ (for distortionless line)} = \sqrt{\frac{Z}{Y}} = \sqrt{\frac{l}{c}} \qquad (1.66)$

(neglecting r and g)

The value of Z_0 depends on the design of the line, and commonly its numerical value in a lossless line comes out to be around 400 Ω. Also, from Eq. (1.32), it is evident that Z_0 is a real number.

Further observation reveals that the voltage and current at the point x can also be denoted as

$$V_x = V e^{j\beta (L-x)} \qquad (1.67)$$

and $\qquad I_x = I_R e^{j\beta (L-x)} \qquad (1.68)$

which clearly shows that at surge impedance loading the power factor is unity. However, the phase angle between the sending-end and the receiving-end quantities is βx rad.

Surge impedance loading (SIL) of a power transmission line is thus the power delivered by a line to a purely resistive load equal to it surge impedance. The line current at SIL is then given by

$$|I_L| = \frac{|V_L|}{\sqrt{3}Z_0} \text{ amp, } V \text{ being the nominal line-to-line voltage.}$$

Also, P_n (i.e. SIL or natural power of the line)

$$= \sqrt{3}|V_L||I_L| = \sqrt{3}|V_L|\frac{|V_L|}{\sqrt{3}|Z_0|}$$

$$= \frac{V^2}{Z_0} \text{ MVA} \qquad (\text{assuming } |V_L| \text{ as } |V|) \qquad (1.69)$$

Power system engineers prefer to express the power capacity of the line (i.e. the *loadability*) in terms of SIL.

The advantage of operating the transmission line at its natural loading is that the voltage profile is flat throughout the length of the line, the insulation stress is uniform at all points and the power factor at all the points along the line is unity. It also means that at SIL, no reactive power is absorbed or generated at either of the two ends of the line. The reactive power generated in the line by the distributed capacitance is exactly consumed by the series inductance of the line. This is a very important observation and utilising this, the reactive power balance for unit length of the line gives

$$V^2 \omega c = I^2 \omega l$$

from which it can also be shown that

$$\frac{V}{I} = \sqrt{\frac{l}{c}} = Z_0$$

This remains true along the entire length of the line. It can be conclusively indicated that the *lossless line achieves the reactive power balance at its natural loading, with a flat voltage profile and unity power factor operation at both ends.*

1.9 REACTIVE LOSS CHARACTERISTICS OF TRANSMISSION LINES

The total series reactive losses in transmission line are given by

$$Q_{SL} = I_L^2 \cdot X$$

$$= \left(\frac{VA}{V}\right)^2 X \qquad (1.70)$$

Noting that for the assumed contingency of constant system load (even with the loss of a heavily loaded EHV line) VA being constant, the change of reactive losses with voltage is obtained from Eq. (1.70) as

$$\frac{dQ_{SL}}{dV} = -\frac{2}{V^3}(VA)^2 \cdot X \qquad (1.71)$$

or from Eq. (1.71) we have

$$\frac{dQ_{SL}}{dV} \propto \left|\frac{1}{V^3}\right| \qquad (1.72)$$

This expression indicates the increase in the series reactive losses with decay in transmission voltage while the OLTC (On Load Tap Changer) at the load-end transformer maintains the system load constant. Unfortunately, the series reactive losses increase inversely with V^3, which in turn depresses the transmission voltage, further enhancing the series reactive losses. Moreover, this effect is further pronounced with reduction in charging capacity (given by $Q_c = CV^2$) of the lines caused by the depression in voltages. If the proper reactive sources to pump reactive power in the system are not available, these mechanisms may lead to a sharp fall in voltage causing voltage collapse.

Equation (1.70) can further be presented in the form of

$$Q_{SL} = (VA)^2 \cdot \left(\frac{1}{V^2/X}\right)$$

$$= \frac{(VA)^2}{SIL} \qquad (1.73)$$

where conventionally SIL, denoting the surge impedance loading, is given by (V^2/Z_0). Figure 1.9 represents the profile of series reactive loss vs. line loading in terms of surge impedance loading. In a real system, the line loading in the post contingency period may be high enough to give rise to a very high series reactive loss, which in turn, depresses the transmission voltage. In extreme cases, these effects may create a high magnitude of line reactive loss for each extra unit of rise in real load. This enormous rise in demand of reactive power causes spontaneous voltage instability of the transmission system and often leaves there little time for operator intervention.

EXAMPLE 1.3 There are four sets of double circuit EHV lines of 400 kV capacity, having a line reactance of 50 ohms per phase for each line and carrying a line current of 1000 amperes. Find how much additional series reactive loss will occur for outage of a single set of line. Assume that the load remains the same.

Solution When all the lines are in service, the series reactive loss (I^2x) is

4 lines × 3 phase × (1000 A)² × 50 ohm = 600 MVAR

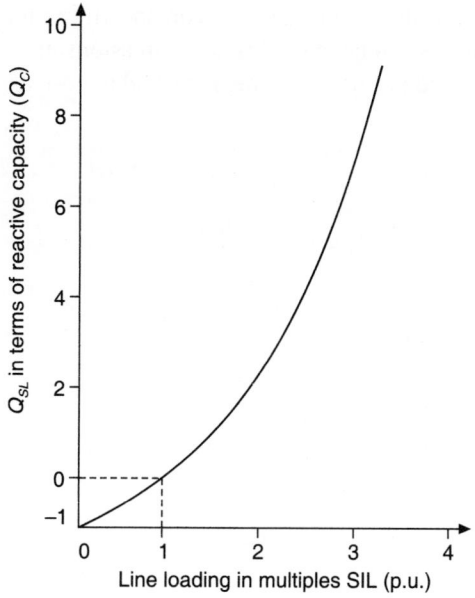

Figure 1.9 Profile of series reactive loss vs SIL.

When one line is out of service, its load is to be catered by the remaining three lines. Hence line current at each 400 kV EHV line is now 1333 A. The new series reactive loss is

$$3 \text{ lines} \times 3 \text{ phase} \times (1333)^2 \times 50 = 800 \text{ MVAR}$$

Hence a single line outage increases the series reactive loss by 33.33%. The effect is voltage drop at the load terminal which further increases the series reactive loss. The reactive power generation is dropped and the situation may be cumulative leading to voltage collapse. Local reactive power reserve is very much necessary to keep the system voltage at or near the nominal value.

1.10 OPERATION OF A TRANSMISSION LINE AT NO-LOAD CONDITION

Assuming the transmission line to be lossless, during no-load operation of the line the receiving-end current is zero and hence Eq. (1.29), for $x = 0$, becomes

$$E = V \cos \beta L = V \cos \theta \qquad (1.74)$$

θ being the line's electrical length in radians. Applying similar reasoning, Eq. (1.30) becomes

$$I_S = j \left[\frac{V}{Z_0} \right] \sin \beta L = j \frac{E}{Z_0} \tan \theta \qquad (1.75)$$

(utilising Eq. (1.74))

The phasor diagram of this condition has been shown in Figure 1.10. It may be convenient to write the voltage and current expressions at any point of the line utilising Eqs. (1.29) and (1.30) under the operation at no-load condition as

$$V_x = V \cos \beta (L - x) = \frac{E \cos \beta (L - x)}{\cos \theta} \quad (1.76)$$

and

$$I_x = j \frac{E}{Z_0} \cdot \frac{\sin \beta (L - x)}{\cos \theta} \quad (1.77)$$

Figure 1.10 No-load phasor diagram of an uncompensated line at no load.

Voltage and current profiles of a typical line at no-load with constant sending-end voltage have been shown in Figures 1.11(a) and (b) respectively at different points of the line across its length.

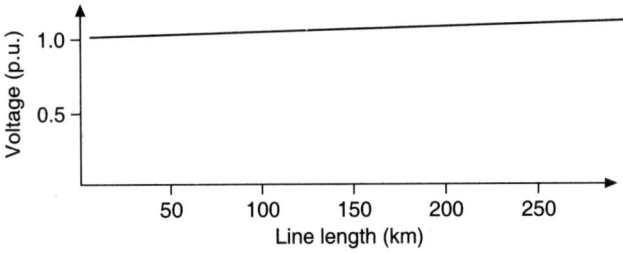

Figure 1.11(a) Voltage profile of an uncompensated line at no load.

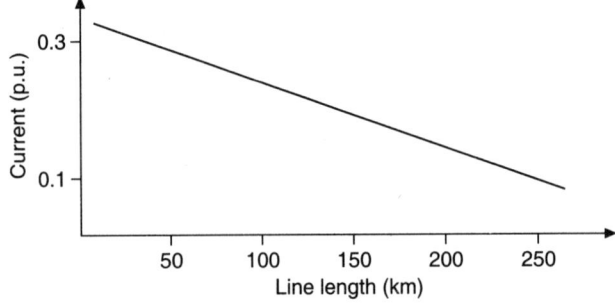

Figure 1.11(b) Current profile of an uncompensated line at no load.

It may be seen that the steady state voltage across the line, operating at no load, increases as the line length increases. This is known as 'Ferranti effect' and may give rise to very high voltages for open-ended or lightly-loaded long lines. This is due to the effect of distributed shunt capacitance along the length of the line.

Ferranti effect can also be found by representing the line's operation in terms of the sending-end voltage and current, and as shown below:

$$E = AV + BI_R \tag{1.78}$$

$$I_S = CV + DI_R \tag{1.79}$$

where

$$A = D = \cos \beta l = \cos \frac{2\pi l}{\lambda} = \cos \frac{2\pi f l}{v} \tag{1.80}$$

where v represent the velocity of electric energy wave and λ the line length. For the open-ended line, $I_R = 0$. This gives from Eq. (1.78)

$$\frac{V}{E} = \frac{1}{A} = \frac{1}{\cos \dfrac{2\pi f l}{v}} \tag{1.81}$$

With an increase in line length, $\left(\cos \dfrac{2\pi f l}{v}\right)$ decreases and hence V increases. In case the line encounters a sudden no-load or load throw at the receiving end, the receiving-end voltage may increase even beyond the calculated value of V given by Eq. (1.76) or (1.81). Immediate measures should be adopted to limit this overvoltage problem.

However, instead of assuming a radial line if the system be assumed to have a synchronous tie, coupling two regional grids, similar reasoning leads to further developments. Taking each of the grids to be equivalent to one synchronous generator, the tie may be assumed to couple two synchronous machines. If E_1 and E_2 be the no-load voltages of the machines,

$$E_1 = E_2 \tag{1.82}$$

Again from Eqs. (1.29) and (1.30), taking $x = 0$,

$$E_1 = E_2 \cos \theta + jZ_0 I_2 \sin \theta \tag{1.83}$$

and

$$I_1 = j\left[\frac{E_2}{Z_0}\right] \sin \theta + I_2 \cos \theta$$

or

$$-I_2 = j\left[\frac{E_2}{Z_0}\right] \sin \theta + I_2 \cos \theta$$

or

$$-I_2 = j\frac{E_2}{Z_0} \frac{\sin \theta}{1 + \cos \theta} = j\frac{E_2}{Z_0} \tan\left(\frac{\theta}{2}\right) \tag{1.84}$$

Substituting the value of I_2 from Eq. (1.84) in Eq. (1.83),

$$E_1 = E_2 \cos\theta + jZ_0 \left(-j\frac{E_2}{Z_0} \tan\frac{\theta}{2}\right) \sin\theta$$

$$= E_2$$

Hence it has been found that for a synchronous tie operation at no load,

$$E_1 = E_2 \tag{1.85}$$

and

$$I_1 = -I_2 = j\frac{E_2}{Z_0} \tan\left(\frac{\theta}{2}\right) \tag{1.86}$$

As E_1 and E_2 are in phase, there will be no power transfer and the current at each end will be the line charging current only. Comparison of Eq. (1.86) with (1.75) reveals that *an uncompensated radial power transmission line is equivalent to two equal halves of synchronous tie lines connected back to back, the charging current of half of the line being supplied from each end* (Figures 1.12(a) and (b)).

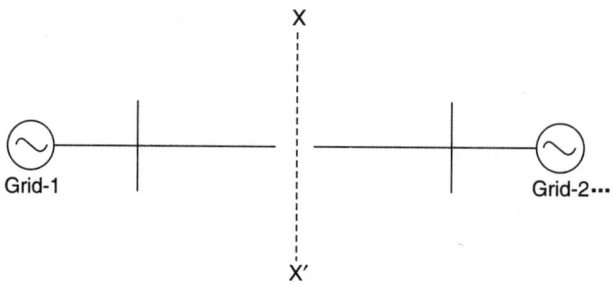

Figure 1.12(a) Synchronous tie line.

Figure 1.12(b) Equivalent representation of a radial line.

The voltage and current at any point for half line is given by

$$V_{(L/2)x} = E_1 \frac{\cos\beta(L/2 - x)}{\cos(\theta/2)} \tag{1.87}$$

and

$$I_{(L/2)x} = j\frac{E_1}{Z_0} \frac{\sin\beta(L/2 - x)}{\cos(\theta/2)} \tag{1.88}$$

It is evident from Eqs. (1.87) and (1.88) that the voltage at the midpoint of the tie (when $x = L/2$) is maximum while the current is minimum. This has been shown graphically in Figures 1.13(a) and (b).

In case the terminal voltage of the tie is not equal at its two ends, the voltage profile is no longer symmetrical, and the point of highest voltage is not at the middle of the line but nearer to the end which has higher system voltage.

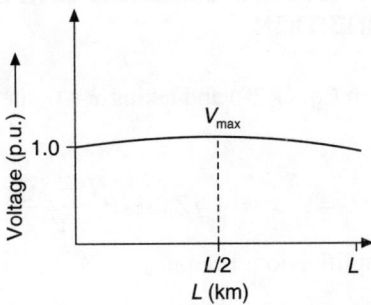

Figure 1.13(a) Rise in voltage at the midpoint of a tie line.

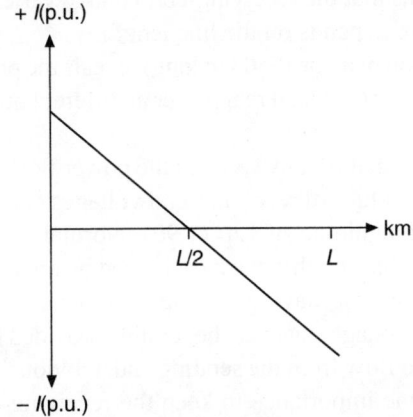

Figure 1.13(b) Decrease in current at the midpoint of an uncompensated line.

Figure 1.13(a) clearly highlights the fact that the imbalance in the reactive power generation and demand raises the voltage at the line, particularly at the midpoint, where the system voltage is totally uncontrolled. The capacitive current at the two ends produces reactive power injection at the terminals. However, the voltage at the two ends, can be kept constant provided the generators at these ends possess the capability of absorption of reactive power. Generally this has been done by setting the generator in the zone of under-excited mode. However, this has got some practical limitations. Firstly, the under-excited mode of operation increases the heating at the stator core of the alternator and secondly, the drop in excitation current reduces the internal induced emf of the alternator thus impairing the stability of the machine. In case the generator is assigned to absorb reactive power more than its limits, this method of controlling the excitation does not serve the purpose and external compensation is strongly recommended.

1.11 OPERATION OF A TRANSMISSION LINE UNDER HEAVY LOADING CONDITION

Utilising $I = \dfrac{P - jQ}{V}$, in Eq. (1.29) and taking $x = 0$, the sending-end voltage is given by

$$E = V \cos\theta + jZ_0 \sin\theta \dfrac{P - jQ}{V} \qquad (1.89)$$

Equation (1.89), on simplification, becomes

$$V = \dfrac{E \pm [E^2 - j2Z_0 \sin 2\theta\,(P - jQ)]^{1/2}}{2\cos\theta} \qquad (1.90)$$

Equation (1.90) reveals that the receiving-end voltage varies with load as well as its power factor. It also depends on the line length.

A typical simulation of a line (350 km long) reveals the profile of the receiving-end voltage plotted against the load real power at different operating power factors (Figure 1.14).

It may be observed that for any value of the power flow below the maximum, there are two possible values of receiving-end voltages corresponding to the two roots of the voltage magnitude of Eq. (1.90). Normal operation of the power system is always desirable and this can be achieved by operating the system at the region of the voltage profile having the higher root of the voltage magnitude. Operation at the lower voltage state may be feasible provided the enormous amount of current is allowed to flow from the sending end. Obviously this is not desirable and hence it is of prime importance to keep the receiving-end voltage near the rated value. The receiving-end voltage reaches the critical state, given by V_{cri} for any load power factor, beyond which voltage instability, denoted by the operation of the power system beyond this critical point, occurs. It is evident from Figure 1.14 that a near flat voltage profile is achieved at unity power factor till the magnitude of critical voltage is attained.

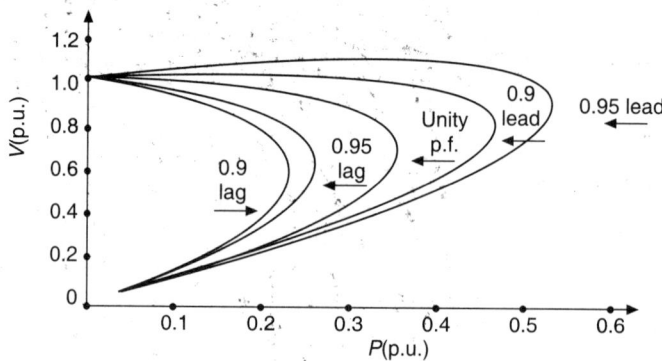

Figure 1.14 Receiving-end voltage profile for varying power flow at different power factors for any specific amount of system reactance.

The load power factor also governs the receiving-end voltage as well as the critical receiving end voltage V_{cri}. Loads with lagging power factor tend to reduce the instantaneous receiving-end voltage as well as the critical receiving-end voltage with increase in power flow. The effect is just the reverse for the leading power factors. Leading power factor loads generate reactive power which supplements the line changing reactive power and in turn enhances the line voltage. The effects of the line length on the critical receiving-end voltage and the instantaneous receiving-end voltage have been shown graphically in Figures 1.15 and 1.16 [6].

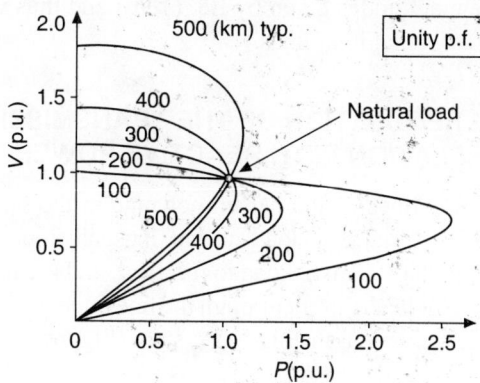

Figure 1.15 Effect of line length on the critical and instantaneous receiving-end voltage at unity power factor.

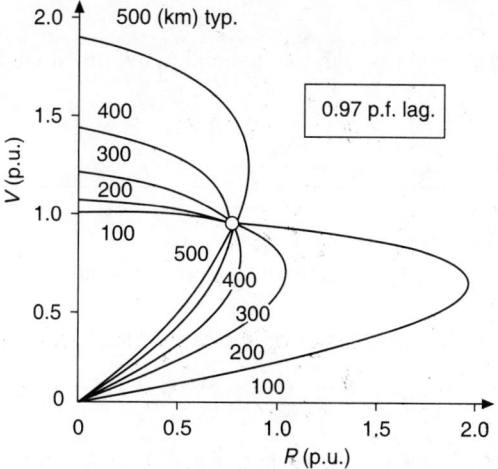

Figure 1.16 Effect of line length on the critical and instantaneous receiving-end voltage at 0.97 p.f. lag.

A simulation revealed that uncompensated lines having lengths between 150 km and 350 km can be operated at normal voltage provided the load power factor is near unity. Operation of radial lines is not feasible at common values of power factor because of the unacceptable voltage profile and lower magnitude of the critical voltage. Its satisfactory operation needs some effective voltage control at the load bus. Even though the voltage magnitudes at the sending and receiving ends are same, at its natural loading, the receiving-end voltage is generally extremely sensitive to any variation in load power beyond the line length of nearly 300 km. The operating voltage state may come down to the vicinity of the lower roots of the voltage magnitudes given by Eq. (1.90) and thus voltage instability occurs particularly for radial long lines.

1.12 VOLTAGE REGULATION OF THE TRANSMISSION LINE AND ITS RELATION WITH REACTIVE POWER

Voltage regulation is defined as the per unit change in the sending-end voltage magnitude for a specific variation in load current. It usually represents the variation in the receiving-end voltage from no load to full load, and is caused by the drop in voltage due to passage of load current through the line impedance. Thus the voltage regulation in p.u. for a simple transmission system (Figure 1.17(a)) is given by

$$\Delta V = \frac{|E|-|V|}{|V|} \quad (1.91)$$

From the phasor diagram (Figure 1.17(b)) corresponding to the system shown in Figure 1.17(b),

$$\Delta V = E - V = IZ_l \quad (1.92)$$

where I is the load current through the line and Z_l the line impedance ($= R_l + jX_l$).

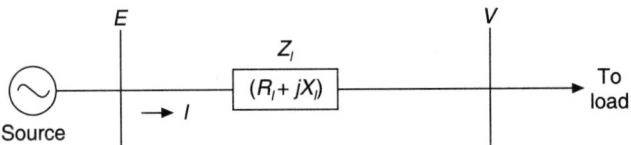

Figure 1.17(a) A simple transmission line model.

Assuming V to be the reference phasor, we can write

$$I^* = \frac{P+jQ}{V} \quad \text{or} \quad I = \frac{P-jQ}{V} \quad (1.93)$$

Substituting the value of I as well as that of Z_l ($= R_l + jX_l$) in Eq. (1.92), we get

$$\Delta V = (R_l + jX_l)\left[\frac{P-jQ}{V}\right]$$

$$= \frac{R_l P + X_l Q}{V} + j\frac{X_l P - R_l Q}{V} \quad (1.94)$$

$$= \Delta V' + j \Delta V'' \quad (1.95)$$

where $\Delta V' = \dfrac{R_l P + X_l Q}{V}$ and $\Delta V'' = \dfrac{X_l P - R_l Q}{V}$

Equation (1.95) reveals that the voltage regulation ΔV has two components $\Delta V'$ and $\Delta V''$, out of which $\Delta V'$ is in phase with V and has been represented by the geometric line 'ab' in Figure 1.17(b), while $\Delta V''$ is in quadrature with V and is represented by the line 'bc' in Figure 1.17(b). It may be noted that the magnitude and phase of V, relative to the sending-end voltage E, are governed by the magnitude and phase of the line current I. This also indicates that the voltage regulation depends on both the real and reactive powers of the load connected at the receiving end.

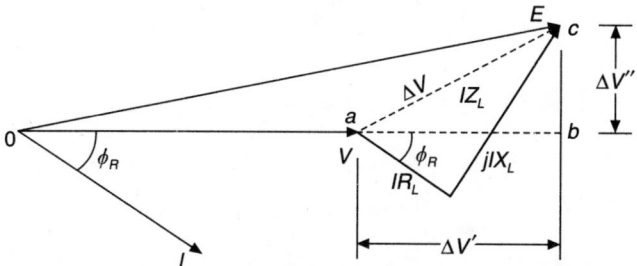

Figure 1.17(b) Phasor diagram of the power system shown in Figure 1.17(a).

A minor alteration in the form of Eq. (1.92) yields the voltage equation for a lossless line as

$$V = E - jIX_l \quad (1.96)$$

Assuming that the power at the sending end equals to that at the receiving-end, I_S the sending-end current (being equal to line current I when distributed line capacitance is neglected) is given by

$$I_S (= I) = \frac{P - jQ}{E} \quad (1.97)$$

From Eqs. (1.97) and (1.96),

$$V = E - j\left(\frac{P - jQ}{E}\right)X_l = E - \frac{X_l}{E}Q - j\frac{X_l}{E}P \quad (1.98)$$

Equation (1.98) reveals that the real power P produces little effect on the receiving-end voltage phasor, since the drop associated with this change is in quadrature with the reference voltage. However, a change in reactive power load Q appreciably affects the receiving-end voltage phasor since the drop associated with this change is in phase with the reference voltage. Thus it may be concluded that *the receiving end voltage is extremely sensitive to any change in reactive power status at the receiving-end bus*.

Further, in power transmission systems, *nodal short-circuit strengths* are frequently termed indicators of robustness. Source reactance being high in weak systems, nodal short-circuit strengths are low and hence these systems are inherently 'weak'. Under the present context, it will be convenient to represent the voltage regulation in terms of short-circuit strength of the system.

The receiving-end short-circuit power, in its complex form, is given by

$$S_{sc} = P_{sc} + jQ_{sc} = VI_{sc}^* = \frac{V^2}{Z_{sc}^*} \qquad (1.99)$$

where

$$Z_{sc} = R_l + jX_l \qquad (1.100)$$

However, assuming $E \simeq V$,

$$R_l = |Z_{sc}| \cos\phi_{sc} = \frac{V^2}{S_{sc}} \cos\phi_{sc} \qquad (1.101)$$

and

$$X_l = |Z_{sc}| \sin\phi_{sc} = \frac{V^2}{S_{sc}} \sin\phi_{sc} \qquad (1.102)$$

where ϕ_{sc} represents the angle between the voltage at the receiving end and the short-circuit current.

$$\phi_{sc} = \tan^{-1} \frac{X_l}{R_l} \qquad (1.103)$$

Substituting the values of R_l and X_l from Eqs. (1.101) and (1.102) in Eq (1.94),

$$\Delta V = \frac{\frac{V^2}{S_{sc}}\cos\phi_{sc} \cdot P + \frac{V^2}{S_{sc}}\sin\phi_{sc} \cdot Q}{V} + j\frac{\frac{V^2}{S_{sc}}\sin\phi_{sc} \cdot P - \frac{V^2}{S_{sc}}\cos\phi_{sc} \cdot Q}{V}$$

or

$$\frac{\Delta V}{V} = \frac{1}{S_{sc}} \left(P\cos\phi_{sc} + Q\sin\phi_{sc} + jP\sin\phi_{sc} - jQ\cos\phi_{sc} \right) \qquad (1.104)$$

But from Eq. (1.95), $\Delta V = \Delta V' + j\Delta V''$

$$\therefore \quad \frac{\Delta V'}{V} = \frac{1}{S_{sc}} \left(P\cos\phi_{sc} + Q\sin\phi_{sc} \right) \qquad (1.105)$$

and

$$\frac{\Delta V''}{V} = \frac{1}{S_{sc}} \left(P\sin\phi_{sc} - Q\cos\phi_{sc} \right) \qquad (1.106)$$

An insight into the above analysis reveals that $\Delta V''/V$ may be ignored as it tends to yield only a phase displacement in the receiving-end voltage with respect to the sending-end voltage. On the other hand, $\Delta V'/V$ is an important component of ΔV that is responsible for altering the magnitude of the voltage at the receiving-end bus.

Equation (1.105) may further be written in the form of

$$\frac{\Delta V'}{V} = \frac{1}{S_{sc}}\left[P\frac{R_l}{Z_{sc}} + Q\frac{X_l}{Z_{sc}}\right] \quad (1.107)$$

In EHV lines, the line resistance is frequently neglected and hence Eq. (1.107) turns out to be

$$\frac{\Delta V'}{V} = \frac{1}{S_{sc}}\left[Q\frac{X_l}{Z_{sc}}\right] = \frac{1}{S_{sc}}[Q \sin \phi_{sc}] \quad (1.108)$$

For small perturbations, Eq. (1.107) can be written as

$$\frac{\Delta V'}{V} = \frac{1}{S_{sc}}\left[\Delta P\frac{R_l}{Z_{sc}} + \Delta Q\frac{X_l}{Z_{sc}}\right]$$

$$= \frac{1}{S_{sc}}\Delta Q \sin \phi_{sc} \quad (1.109)$$

(neglecting R_l)

ϕ_{sc} being large, $\sin \phi_{sc} = 1$. Hence Eq. (1.109) finally becomes

$$\frac{\Delta V}{V} = \frac{\Delta V'}{V} = \frac{1}{S_{sc}}\Delta Q \sin \phi_{sc} = \frac{\Delta Q}{S_{sc}} \quad (1.110)$$

Thus, in a lossless transmission system, *the per unit voltage change in the receiving-end bus is equal to the ratio of change in reactive power status to the short-circuit strength at that bus.*

Further, from Eq. (1.108)

$$\frac{\Delta V'}{V} \simeq \frac{\Delta V}{V} = \frac{1}{S_{sc}} \cdot Q$$

where $\sin \phi_{sc} \simeq 1$.

Now,

$$\frac{E - V}{V} = \frac{Q}{S_{sc}}$$

$\therefore$
$$V = \frac{E}{1 + Q/S_{sc}} \simeq E\left[1 - \frac{Q}{S_{sc}}\right] \text{ assuming } Q \ll S_{sc} \quad (1.111)$$

Equation (1.111) represents the approximate relationship between the voltage at receiving end and the reactive power in terms of short-circuit strength. Figure 1.18 represents the graphical view of this result, the gradient of the load line being represented by $(-E/S_{sc})$

EXAMPLE 1.4 A 100 km, 400 kV transmission line has system reactance of 0.25 ohms/km per phase and source reactance of 25% of line reactance. Find the supply voltage/phase when the load side p.f. is (a) unity, (b) 0.8 lag. Assume a power transfer of 800 MW.

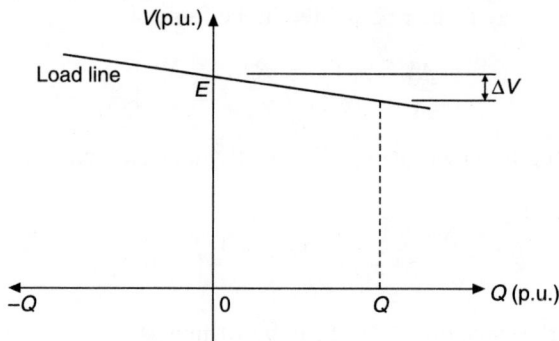

Figure 1.18 Profile of voltage vs receiving-end reactive power.

Solution From the given data,

$$V = \frac{400}{\sqrt{3}} = 231 \text{ kV}$$

$$x_{\text{line}} = 0.25 \times 100 = 25 \text{ ohm}$$

$$x_{\text{source}} = \frac{1}{4} \times 0.25 \times 100 = 6.25 \text{ ohm}$$

(a) Since in the first case, $\cos \phi$ (load) = 1.0

$\therefore \quad Q_r = 0$

Hence $\quad E_{\text{ph}} = V + jI(x_l + x_s)$

$\qquad = 231 \times 10^3 + j1154.73(25 + 6.25)$

$$\left[\because I = \frac{800 \times 10^6}{\sqrt{3} \times 400 \times 10^3} = 1154.73 \text{ A} \right]$$

$\qquad = 231 \times 10^3 + j1154.73 \times 31.25$

$\qquad = 231 + j36.084$

$\qquad = 233.81 \angle 8.8° \text{ kV}$

$\therefore \quad |E|_{L-L} = 233.81 \times \sqrt{3} = 404.96 \text{ kV}$

(b) In the second case, $\cos \phi$ (load) = 0.8

$\therefore \quad Q_r = P_r \tan \phi$

$\qquad = 600 \text{ MVAR}$

From $\quad E_{\text{ph}} = V + \dfrac{X Q_r}{V} + j\dfrac{X P_r}{V}$, we get

$$E_{\text{ph}} = 231 \times 10^3 + \left[\frac{25 + 6.25}{231 \times 10^3} \times 600 + j \frac{25 + 6.25}{231 \times 10^3} \times 800 \right]$$

$\qquad = 231 \times 10^3 + 81.17 \times 10^3 + j108.22 \times 10^3$

$$= 330.396 \angle 19.12° \text{ kV}$$
$$\therefore \quad |E|_{L-L} = 572.27 \angle 19.12° \text{ kV}$$

Hence it is evident that at unity p.f. of load side, the source-side voltage is to be enhanced approximately by 1.2% while for the receiving end p.f. of 0.8 lag, the source-side voltage is to be enhanced by 43%. The power angle also increases though the change is nominal.

EXAMPLE 1.5 Find the capacity of a static VAR compensator to be installed at a bus with ±5% voltage fluctuation. The short-circuit capacity is 5000 MVA.

Solution For the switching of static shunt compensator, let

ΔV = voltage fluctuation
ΔQ = reactive power variation
  (i.e. the size of the compensator)
S_{sc} = system short-circuit capacity

Since
$$\Delta V = \frac{\Delta Q}{S_{sc}},$$
or
$$\Delta Q = \Delta V \times S_{sc},$$
we get
$$\Delta Q = \pm (0.05 \times 5000)$$
$$= \pm 250 \text{ MVAR}$$

The capacity of the static VAR compensator is ±250 MVAR.

EXAMPLE 1.6 In a radial transmission system, the p.u. values are referred to 100 MVA base and the voltage bases as shown in Figure E1.3. Find the p.f. of the generator at which it should operate. (Assume $V = 1.0$ p.u.)

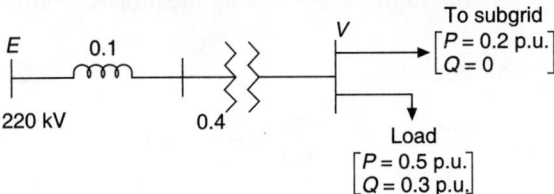

Figure E1.3 A radial transmission system.

Solution At the load bus,
$$P = 0.5 + 0.2 = 0.7 \text{ p.u.}$$
$$Q = 0.3 + 0 = 0.3 \text{ p.u.}$$

Total $I^2 X$ loss in transformer at receiving bus and line

$$= \frac{P^2 + Q^2}{V^2} \cdot (X_{TR} + X_{line})$$

$$= \frac{0.7^2 + 0.3^2}{1^2} \times 0.5 = 0.29 \text{ p.u.}$$

$\therefore$ Net $Q = 0.3 + 0.29 = 0.59$ p.u. (demand)
Since $P = 0.7$ p.u. and $Q = P \tan \phi$, hence

$$\phi = \tan^{-1} \frac{Q}{P} = \tan^{-1} 0.8428 = 40.13°$$

$\therefore$ pf (required) $= \cos 40.13° = 0.765$

The generator operates at 0.765 pf (lag).

1.13 MAXIMUM POWER TRANSFER IN AN UNCOMPENSATED LINE

Rearranging Eq. (1.59), the power flow in any line is given by

$$P = \frac{EV \sin \delta}{Z_0 \sin \theta} \qquad (1.112)$$

Equation (1.112) is identical to the conventional power flow expression $\left(P = \frac{EV}{X_l} \sin \delta\right)$ and is valued for synchronous loads as well as asynchronous loads, and also where $E \neq V$.

But $\qquad Z_0 = \sqrt{\frac{l}{c}} \quad \text{and} \quad \theta = \beta L = \omega L \sqrt{lc}$

$\therefore \qquad Z_0 \sin \theta = Z_0 \theta = \omega L l = X_l$

$$\left[\because \beta = \omega \sqrt{lc} = \frac{\omega}{v}, v \text{ being the propagation velocity} = \frac{1}{\sqrt{lc}}\right]$$

Equation (1.112) now becomes

$$P = \frac{EV \sin \delta}{Z_0 \sin \theta} = \frac{EV}{X_l} \sin \delta$$

βL becomes $\frac{\omega}{v} L = \frac{2\pi f L}{v} = 0.06 L$ (for a 50 Hz system),

$$\left[\because \beta = \frac{2\pi}{v/f} = \frac{360}{3 \times 10^5 / 50} = 0.06\right]$$

If the terminal voltages are fixed, the power transmitted is a function of power angle δ. If $E = V = V_0$ (say), Eq. (1.112) can be represented in the form of

$$P = \frac{V_0^2}{Z_0} \cdot \frac{\sin \delta}{\sin \theta} = \frac{P_x}{\sin \theta} \sin \delta \qquad (1.113)$$

while V_0^2/Z_0 represents the SIL, i.e. P_x. Figure 1.19 represents the profile of P vs δ. It states that successful power transfer can be done till δ reaches 90°, beyond $\delta = 90°$, the system stability is affected. $\delta = 90°$ represents the *steady state stability limit* and the power transfer is maximum at this point. The steady state stability limit is then given by

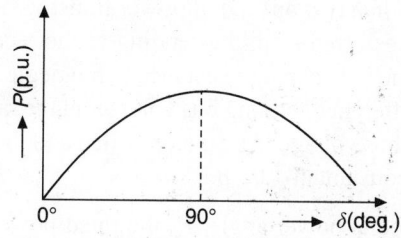

Figure 1.19 $P-\delta$ profile of a power line.

$$P_{max} = \frac{EV}{Z_0 \sin\theta} = \frac{P_x}{\sin\theta} \quad (\text{where } E = V) \qquad (1.114)$$

It may be mentioned at this stage that with enhancement of line lengths, θ increases and hence the steady-state stability limit decreases. Table 1.3 represents the effect of enhancement of line length on steady-state stability limit.

Table 1.3 Effect of enhancement of line length on steady-state stability limit

Line length km	$\theta \text{ (deg)} \left[= \frac{2\pi}{v/f} \times L \right]$	P_{max}/P_0 (p.u.)
100	6	9.57
200	12	4.81
300	18	3.24
400	24	2.46
500	30	2.00
600	36	1.70
800	48	1.35

An uncompensated line is seldom operated near its steady-state stability limit because this may lead the system to the zone of system instability, once the system encounters any occasional or major disturbance caused by faults/switching operations/sudden load variations, etc. It is customary to keep the system operating angle much below the steady state stability limit (the power angle usually does not exceed 30°). However, this restricts the magnitude of the transmissable power at half the maximum amount of power that can be transmitted. It is also evident from the above analysis that smaller power can be transmitted over longer distances and once the line length increases, the steady-state stability limit itself drops.

Moreover, with enhancement of line loading, the system needs reactive power support such that the voltage profile can be maintained. Hence, it can be concluded that medium and long lines face limitations in the capability of transporting real power due to stability problems. Also, it may be extremely difficult to maintain the rated voltages at the two ends of the line if the line is not properly compensated.

EXAMPLE 1.7 In a two bus radial power transmission system, 0.8 p.u. of real power is being transferred from the sending-to the receiving-end bus. If the p.u. reactance of the line is 1.00 p.u., resistance being neglected, plot the profile of the power angle for different operating bus voltages at the rated power level, assuming the receiving-end bus voltage to be uncontrolled and the voltage magnitude of the sending-end bus constant at 1.00 p.u.

Solution If δ be the power angle, P_0 the rated power flow, E the sending-end voltage and V the receiving end voltage and x the reactance of the line, then

$$\sin \delta = \frac{P_0 x}{EV}$$

Given, $\qquad P_0 = 0.8, \quad x = 1.0, \quad E = 1$

When, $\qquad V = 0.8, \quad \sin \delta = \dfrac{0.8 \times 1}{1 \times 0.8} = 1$

or $\qquad \delta = 90°$.

When, $\qquad V = 0.9, \quad \sin \delta = \dfrac{0.8 \times 1}{1 \times 0.9} = 0.888$

or $\qquad \delta = 62.62°$.

When, $\qquad V = 1.0, \quad \sin \delta = \dfrac{0.8 \times 1}{1 \times 1} = 0.8$

or $\qquad \delta = 53.13°$.

When, $\qquad V = 1.1, \quad \sin \delta = \dfrac{0.8 \times 1}{1 \times 1.1} = 0.727$

or $\qquad \delta = 46.65°$.

When, $\qquad V = 1.2, \quad \sin \delta = \dfrac{0.8 \times 1}{1 \times 1.2} = 0.667$

or $\qquad \delta = 41.81°$.

When, $\qquad V = 1.3, \quad \sin \delta = \dfrac{0.8 \times 1}{1 \times 1.3} = 0.615$

or $\qquad \delta = 37.98°$

Table 1.4

V (p.u.)	δ (in degrees)
0.8	90
0.9	62.73
1.0	53.13
1.1	46.65
1.2	41.81
1.3	37.98

Figure E1.4 represents the profile of δ vs. V for power flow of 0.8 p.u. of power and at different receiving-end voltages. This clearly highlights that with enhancement of bus voltage magnitude at the receiving end, power angle decreases, i.e. the steady-state stability margin improves.

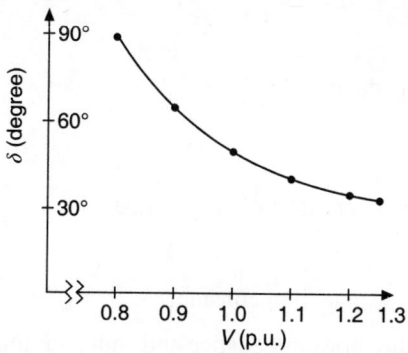

Figure E1.4

1.14 LINE LOADABILITY

For an EHV transmission system, the power transfer capability of transmission links is substantially affected by nodal power injections and topological changes. It has been generally accepted that the nodal strength, i.e. the *short-circuit capacity*, is the *nodal indicator of robustness* of any power network. It is the inverse of the positive sequence equivalent impedance in p.u., when seen for the node considered. Frequently, this equivalent impedance (hereinafter will be called simply as source reactance) is dominated by the series reactance of the transmission line as well as terminal reactance when analysed in a lossless frame.

A simple form of power supply system, being represented by an equivalent π network is shown in Figure 1.20. The expression for line loadability (S) in per unit (p.u.) quantities can be derived in terms of the surge impedance loading (SIL) and is given below.

$$S_0(\text{SIL}) = \frac{V^2}{Z_0} \tag{1.115}$$

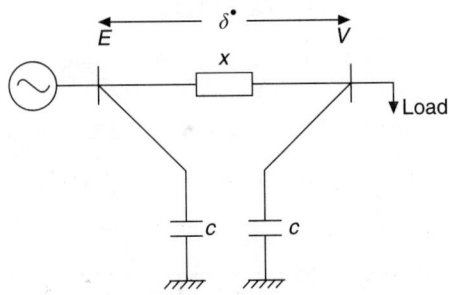

Figure 1.20 Schematic of a simple power line model.

When the system voltage is held at its nominal value in per unit, Eq. (1.115) can be written as

$$S_0 = \frac{1}{Z_0} \quad (1.116)$$

Z_0 being the surge impedance.
Also,

$$Z_c \text{ (the characteristic or surge impedance)} = \left(\frac{Z_p}{Y_p}\right)^{1/2} \quad (1.117)$$

$$\gamma \text{ (the propagation constant)} = \alpha = j\beta = \sqrt{Z_p Y_p} \quad (1.118)$$

where, Z_p and Y_p are the series impedance and shunt admittance per phase, α the attenuation constant and β the phase constant.

In the basic power system, the receiving-end power (S_R) is given by

$$S_R = P_R + jQ_R = VI^* \quad (1.119)$$

where

$$I = \frac{E-V}{Z} - \frac{Y}{2}V$$

$$= 2E - 2V - ZYV/2Z \quad (1.120)$$

When all the quantities are in per unit, the receiving-end power may be expressed as

$$\frac{S_R}{SIL} = \frac{VI^*}{(1/Z_0)} = (Z_0)\left[\frac{2E - 2V - ZYV}{2Z}\right]^* V$$

$$= Z_0 \left[\frac{2(E/V) - 2 - (Z_0 \sinh \gamma L)(2/Z_0 \tanh \gamma L/2)}{2Z_0 \sinh \gamma L}\right]^* |V|^2$$

$$= \frac{Z_0}{Z_0^*}\left(\frac{E/V - \cosh \gamma L}{. \sinh \gamma L}\right)^* |V|^2 \quad (1.121)$$

For the given values of E, V and L (line length), (S_R/SIL) is the function of the propagation constant γ and the ratio (Z_0/Z_0^*). Because the series resistance R of the EHV transmission lines is normally much smaller than its inductive reactance X, the attenuation constant α is very small, and the propagation constant γ approaches the phase constant $(j\beta)$. Surge impedance being a real quantity, $Z_0^* = Z_0$.

Equation (1.121) then reduces to

$$\frac{S_R}{\text{SIL}} = j\frac{(E/V)^* - \cos\beta L}{\sin\beta L}|V|^2$$

where
$$\left.\begin{array}{l}\beta = \dfrac{\omega}{v}, \\ \omega = 2\pi f\end{array}\right\} \quad (1.122)$$

v being the velocity of light.

It is interesting to note that the line loadability, defined as *the magnitude of receiving-end power under a certain operating condition of the line*, is independent of the line's electrical parameters. It depends exclusively on the line length and terminal voltages along with angular separation between them. S_R, in terms of SIL, represents line loadability.

Simulations for determining and comparing the line loadability of a set of 'robust' and 'weak' systems have been graphically illustrated in Figure 1.21 from which it is clear that the 'line loadability' of robust systems is better compared to that of weak capacity systems.

High source reactance of EHV transmission systems plays a vital role in limiting the line loadability as evident from Figure 1.21. Loadability can be improved by relaxing the voltage drop constraint for small and medium lines

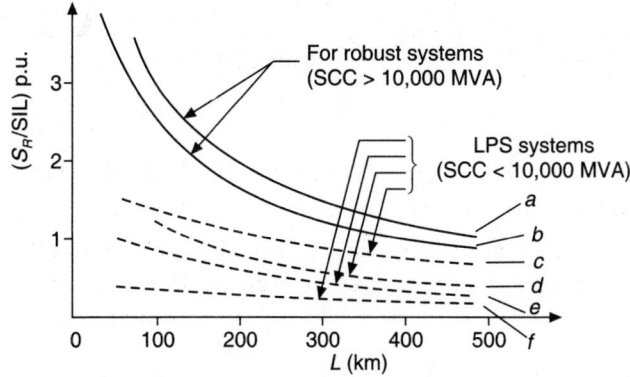

Figure 1.21 Line loadabilities of "strong" and "weak" systems
a,b—Strong systems
c, d, e, f—Weak systems
SCC—Short Circuit Capacity.

only, as well as by adopting conventional methods for reduction in line power loss. Series capacitor insertion also enhances the loadability. The simulation also exhibits the impairment of line loadability for shunt reactor compensated systems. However, this degree of this impairment is low for robust systems.

EXERCISES

1. Name the primary constants of the model of a transmission line. How are they related to the secondary constants?
2. Briefly state the concepts of lumped parameter models of a transmission line. What are the equations for sending end voltage and current at the sending and at the receiving ends of a line?
3. Develop expressions for power flows at the sending and receiving-ends of a power line.
4. How do you obtain the expressions for the receiving-end of a power circle diagram? How can you find the expression of maximum power from it?
5. Obtain the expression for reactive power requirements of a transmission line.
6. What is surge impedance loading? What is its expression?
7. Obtain the reactive loss characteristics of a transmission line.
8. How would you model the operation of an uncomponsated line at no load? What is Ferranti effect?
9. Find an expression for the receiving-end voltage in a heavily-loaded uncompensated line. Draw the relevant characteristics.
10. Establish relationships between the voltage regulation of a short line model and reactive power. How is it governed by system short-circuit capacity?
11. How do you obtain a relation for maximum power and line angle/line length of a power line?
12. What is line loadability? Obtain an expression for it.

Chapter 2

Reactive Power Flow and Voltage Control Problems

2.1 INTRODUCTION

We have seen earlier that the voltage regulation in a transmission line is very much dependent on the reactive power status of the load bus. In fact, the voltage control problems generate from the reactive power mismatch problem, i.e. the condition when the reactive power generation is either lower or higher than the reactive power demand. During the heavy loading condition, the reactive power demand is high leading to higher voltage regulation while during the light loading condition of a medium or long line the reactive charging is capacitive type and may lead to surplus reactive generation. It may result in increase in receiving-end voltage (i.e. Ferranti effect). In this chapter, we will discuss different aspects of reactive power conditions and the consequent voltage control problems.

2.2 REACTIVE POWER-VOLTAGE (Q-V) COUPLING CONCEPT

The well-known real and reactive power equations at the sending and receiving ends of a basic power transmission line model are given by

$$P_S = \frac{EV}{X} \sin \delta \tag{2.1}$$

$$Q_S = \frac{E^2}{X} - \frac{EV}{X} \cos \delta \tag{2.2}$$

$$P_R = \frac{EV}{X} \sin \delta \tag{2.3}$$

$$Q_R = \frac{EV}{X}\cos\delta - \frac{V^2}{X} \tag{2.4}$$

With $\delta = 90°$ (at steady-state stability limit),

$$P_S = P_R = P_{max} = \frac{EV}{X}$$

The active power Eq. (2.1) or (2.3) can be written as

$$P_S = P_R = P_{max}\sin\delta \tag{2.5}$$

while the reactive power Eq. (2.2) or (2.4) can be written as

$$Q_S = -Q_R = P_{max}(1-\cos\delta) \tag{2.6}$$

[with the assumption that $E = V$]

Figure 2.1 exhibits the profiles of P_R, Q_R with respect to the power angle.

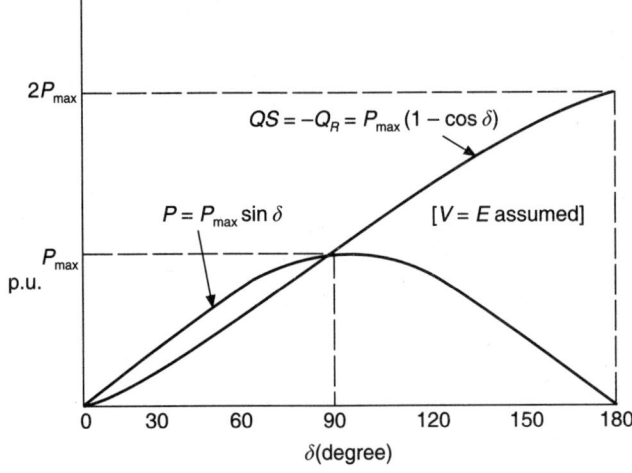

Figure 2.1 Profiles of P_R, Q_R with respect to the power angle.

We are interested in the lower power angle operation. With δ assumed to be small $\sin\delta \to \delta$ while $\cos\delta \to 1$. Thus the sending-end and the receiving-end power expressions are modified as

$$P_S = P_R = P_{max}\delta \tag{2.7a}$$

$$Q_S = -Q_R = \frac{E^2}{X} - \frac{EV}{X} \tag{2.7b}$$

We thus conclude here (in an approximated form),

- *P and δ are closely coupled*, while
- *Q and V are closely coupled*.

These physical relationships are reasonably applicable with acceptable accuracy as we have seen their application in *Fast Decoupled Load Flow* (*FDLF*) algorithm in load flow studies.

2.3 GOVERNING EFFECTS ON REACTIVE POWER FLOW

2.3.1 Line Demand

Let us take a simple two bus power system with $E = 1$ p.u., $V = 0.95$ p.u. and $\delta = 30°$.

$$Q_S = \frac{1^2 - 1 \times 0.95 \times 0.866}{X} = \frac{0.178}{X}$$

while

$$Q_R = \frac{1 \times 0.95 \times 0.866 - 0.95^2}{X} = -\frac{0.081}{X}$$

The negative value of Q_R indicates that the line is demanding reactive power from the receiving end. The transmission line has thus become **drain** of reactive power, absorbing reactive power from the sending end and demanding reactive power from the receiving-end; the line reactive loss is becoming $\frac{(0.178 + 0.081)}{X}$, i.e. $\frac{0.26}{X}$ p.u.

Thus for a system with depressed receiving-end voltage, the line demands reactive power.

2.3.2 Effect of Receiving-end Bus Loading

Next, we look at the receiving-end power circle diagram (Figure 2.2) for a 400 kV, 150 km, 100 MVA power system model.

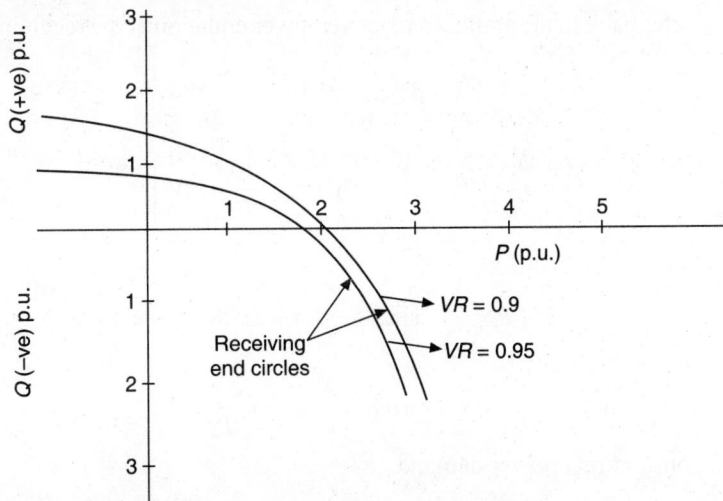

Figure 2.2 Receiving-end power circle diagram.

We see that for higher real power transfer, the receiving-end reactive power becomes negative indicating that the line is demanding reactive power from the receiving-end. With drop in receiving-end voltage, the situation improves marginally.

- *With higher real loading, the reactive demand increases heavily.*

2.3.3 Implication of Voltage Regulation

We can express the voltage regulation of a transmission line as
$$(E - V) = V_i + jV_q$$
where $V_i = (I_p R + I_q X)$ and $V_q = (I_p X - I_q R)$

(the suffixes i and q indicate the in-phase and quadrature components). Neglecting the resistance of the line,
$$V_i = I_q X \quad \text{while} \quad V_q = I_p X$$
i.e.
$$(E - V) = (I_q VX)/V + j(I_p VX)/V$$
$$= Q_R X/V + jP_R X/V \qquad (2.8)$$

This clearly shows that:

- *Voltage regulation increases with higher power transfer*
- *Angle between V_i and V_q enhances when voltage regulation is higher.*

Since with higher real power transfer the demand for reactive power increases, it becomes evident that *for heavily loaded lines the voltage regulation increases and becomes very sensitive to reactive demand Q_R.*

2.3.4 Implication of Magnitude of Power Angle

Again we refer back to the simlified reactive power equation at the receiving end,
$$Q_R = \frac{EV}{X} \cos\delta - \frac{V^2}{X}$$
$$\frac{dQ}{dV} = \frac{1}{X}\left[E\cos\delta - EV\sin\delta \frac{d\delta}{dE} - 2V \right] \qquad (2.9)$$

But for
$$P = \frac{EV}{X}\sin\delta$$
$$\frac{dP}{dE} = \frac{V}{X}\sin\delta + \frac{EV}{X}\cos\delta \frac{d\delta}{dE} \qquad (2.10)$$

and with constant real power demand,
$$\frac{dP}{dE} = 0$$

Then from Eq. (2.10),
$$\frac{d\delta}{dE} = -\frac{\tan\delta}{E}$$

Finally from Eq. (2.9),
$$\frac{dQ}{dV} = \frac{1}{X}\left[E\cos\delta - EV\sin\delta\left(-\frac{\tan\delta}{V}\right) - 2V\right]$$

$$= \frac{1}{X}\left[\frac{E}{\cos\delta} - 2V\right] \quad (2.11)$$

With high values of δ, $\cos\delta \to 0$ and $dQ/dV \to \infty$.

- Then it becomes evident that at high values of power angle dV/dQ approaches 0 indicating loss of voltage at load bus.

2.3.5 Load Bus Reactive Power Sensitivity

From load flow equations, we can write

$$\begin{bmatrix}\Delta P\\ \Delta Q\end{bmatrix} = \begin{bmatrix}\frac{\partial P}{\partial \delta} & \frac{\partial P}{\partial V}\\ \frac{\partial Q}{\partial \delta} & \frac{\partial Q}{\partial V}\end{bmatrix}\begin{bmatrix}\Delta\delta\\ \Delta V\end{bmatrix} = [J]\begin{bmatrix}\Delta\delta\\ \Delta V\end{bmatrix} \quad (2.12)$$

With analogous reasoning of FDLF model (i.e. the decoupling concept), it is considered here that

$$\frac{\partial P}{\partial V} = 0 \quad \text{and} \quad \frac{\partial Q}{\partial \delta} = 0$$

Thus,
$$[\Delta Q] = \left[\frac{\partial Q}{\partial V}\right][\Delta V] = [J][\Delta V] \quad (2.13)$$

At load buses, $\delta Q_L / \delta V_L$ (diagonal elements of J) represents the *reactive power sensitivity*.

- Obviously $\delta Q_L / \delta V_L$ approaching 0 or ∞. It indicates instability and load bus voltage remains stable with $\delta Q_L / \delta V_L$ remaining at positive values.

The degree of stability can also be assumed by observing the magnitude of $(\delta Q_L / \delta V_L)$.

2.3.6 Effect of Series Reactive Loss

The *series reactive loss* is given by
$$|Q_L| = |I^2 X|$$

For constant MVA load,

$$|I| = \left|\frac{S}{V}\right|$$

Therefore,
$$Q_L = \left|\frac{S}{V}\right|^2 |X|^2 \qquad (2.14)$$

or the change in series reactive loss for a change in voltage is given by

$$\frac{dQ_L}{dV} = -\frac{2}{V^3} |S|^2 |X| \qquad (2.15)$$

Assuming $|S|$ and $|X|$ to be constant,

$$\left|\frac{dQ_L}{dV}\right| \propto \frac{1}{|V|^3} \qquad (2.16)$$

i.e. the change in series reactive loss increases sharply with any drop in transmission voltage.

We thus conclude the following:

- Q and V are closely coupled.
- For heavily loaded lines and with depressed receiving-end voltage, the line acts as a *drain* of reactive power.
- With higher level of real power loading, reactive demand of the system increases sharply and may cause voltage reduction at the receiving-end bus with no additional support of reactive power.
- Line voltage regulation increases with higher amount of power transfer.
- Voltage regulation is particularly sensitive to reactive power demand of the system.
- At higher values of power angle, there is a chance of loss of voltage at the load bus.
- Diagonal elements of Jacobian indicate reactive sensitivity at load bus and for stability of load bus voltage, this sensitivity should be positive (its value of zero or infinity indicates instability).
- Series reactive loss is voltage dependent and the change in series reactive loss in the line would sharply increase with any decline in receiving-end voltage.

EXAMPLE 2.1 In a two-bus system the real load at the receiving end is zero. Obtain the expression for critical receiving-end voltage and the sensitivity (dQ_S/dQ) when the receiving end is loaded with purely inductive loads. Comment on your findings.

Solution The reactive power equation at the receiving-end bus is given by

$$Q = \frac{EV}{X} \cos\delta - \frac{V^2}{X}.$$

In this problem, it is given that $P = 0$ ($\because$ there is no real power demand at the receiving-end bus).
This gives $\delta = 0$

$\therefore$
$$Q = \frac{EV}{X} - \frac{V^2}{X} \qquad \text{(i)}$$

and
$$\frac{dQ}{dV} = \frac{1}{X}(E - 2V) \qquad \text{(ii)}$$

At voltage stability limit, $\frac{dQ}{dV} = 0$;

this gives the critical receiving-end voltage, obtained from (ii) with $\frac{dQ}{dV} = 0$, to be

$$V_{cri} = \frac{E}{2}.$$

$\therefore$ Max reactive power expression is given by

$$Q_{max} = \frac{2V_{cri}^2}{X} - \frac{V_{cri}^2}{X} = \frac{V_{cri}^2}{X} = \frac{E^2}{4X}$$

However, the sending-end reactive power is given by

$$Q_S = Q + \text{Line reactive power loss}$$

i.e.
$$Q_S = Q + I^2 X = Q + \frac{Q_S^2 X}{E^2} \qquad \text{(iii)}$$

$$\left[\because I = \frac{\text{sending-end complex power }(S)}{\text{sending-end voltage }(E)} = \frac{Q_S}{E}\right.$$

$$\text{as sending-end complex power } S \text{ turns out to be}$$
$$\left. Q_S \text{ in the absence of any real power demand}\right]$$

i.e. from (iii), we can write

$$Q_S^2 - \frac{E^2}{X} Q_S + \frac{E^2}{X} Q = 0$$

or
$$Q_S = \frac{1}{2}\left[\frac{E^2}{X} \pm \sqrt{\frac{E^4}{X^2} - \frac{4QE^2}{X}}\right]$$

and
$$\frac{dQ_S}{dQ} = \frac{1}{\left(1 - \frac{4 \times Q}{E^2}\right)^{1/2}} = \frac{1}{\left(1 - \frac{Q}{Q_{max}}\right)^{1/2}}$$

Thus, we see that with no load at the load bus ($P = 0$, $Q = 0$), the receiving-end voltage approaches E, and (dQ_S/dQ) becomes unity. On the other hand, with purely reactive load at load bus, (at $Q = Q_{max}$), $V = \frac{E}{2}$ while (dQ_S/dQ) = ∞. This is a clear case of voltage collapse and we can infer that with large amounts of reactive power demand, the sending-end system gets stressed and (dQ_S/dQ) approaches a very high value. A lower value of (dQ_S/dQ) is thus always preferred.

2.4 ASPECTS OF REAL AND REACTIVE POWER STATIC STABILITY

2.4.1 Steady State (Static) Real Power Stability

A preliminary discussion has been presented in the earlier chapter regarding the steady stability limit of a power line. However, to extend the understanding of the concept, further discussion is needed. Here, a two-machine example is considered with no shunt branch taken into account. For simplicity, the system will be assumed to be lossless.

Thus, the analytical expression for real power flow becomes (same as Eq. (1.50)).

$$P_1 = V_1 V_2 Y_{12} \sin \delta_{12} \quad (2.17)$$

and
$$P_2 = -V_1 V_2 Y_{12} \sin \delta_{12} \quad (2.18)$$

while the reactive power flow is analytically represented by the second part of Eq. (1.50) and is reproduced here as

$$Q_1 = V_1^2 Y_{12} - V_1 V_2 Y_{12} \cos \delta_{12} \quad (2.19)$$

and
$$Q_2 = V_1 V_2 Y_{12} \cos \delta_{12} - V_2^2 Y_{12} \quad (2.20)$$

Setting all the derivatives of Eqs. (2.17) to (2.20) to zero, a set of nonlinear algebraic equations is obtained. Solution of these equations leads to the mathematical representation of 'equilibrium' point. For a particular system when Y_{12} is known, the equilibrium point will depend on a given set of performance parameters (i.e. P_1, Q_1, P_2 and Q_2). However, in addition to these parameters, there is another set of free variables, viz. ($V_1 - V_2$) and ($\delta_1 - \delta_2$). Hence there may not be a single solution of the power flow equations for any arbitrary values of these parameters and the free variables.

Assuming fixed bus voltages, the P-δ characteristic of a power transmission system can be plotted at the receiving end and is shown in Figure 2.3. It may be observed that there is no load flow solution beyond $P_2 = P_{2\max}$. Below $P_2 = P_{2\max}$, there are always two equilibrium points (i.e. two-load flow solutions). Out of these positions, the state of operation corresponding to parts a, b, c, etc. below point '0' is said to be stable (†) while the state of operation corresponding to points a', b', c' and any other point beyond point '0' is said to be unstable[†]. Critical state of stability is achieved at '0' (which is also termed steady-state stability limit). In the zone of normal operation, i.e. till $P_2 = P_{2\max}$, any enhancement in power angle (which occurs due to increase in power demand) causes enhancement in power transfer, while beyond point '0' in the P-δ curve of Figure 2.3, for any increase in system's real power demand, the power transfer drops.

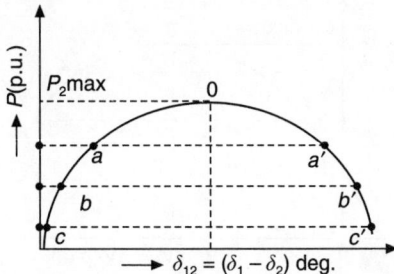

Figure 2.3 P-δ characteristic to explain steady state stability of real power.

2.4.2 Steady State (Static) Reactive Power Stability

Reactive power stability shares all aspects of real power stability. However, there is one important difference. While the real power expression that gives the basic

---

[†] As $P = f(V, \delta)$, hence for small perturbations,

$$\Delta P = \frac{\partial P}{\partial \delta} \Delta \delta + \frac{\partial P}{\partial V} \Delta V$$

Applying the concept of decoupled load flow [1, 35],

$$\frac{\partial P}{\partial V} = 0$$

Therefore,

$$\Delta P = \frac{\partial P}{\partial \delta} \Delta \delta = C \Delta \delta \quad \text{where } C = \frac{\partial P}{\partial \delta} \quad \text{(say)}$$

i.e.
$$\Delta \delta = \frac{\Delta P}{C}$$

and as C approaches zero, $\Delta \delta$ approaches infinity.
Hence $\Delta \delta$ is finite and positive iff $C > 0$. $\Delta \delta$ becomes negative when $C < 0$ and the critical condition is achieved when $C = 0$. Thus system stability can only be achieved when $C > 0$; critical state of stability is at $C = 0$ and the system lands to an unstable state when $C < 0$.

building block of real power steady state stability criterion is trigonometric Eq. (2.17) or (2.19), the same for reactive power stability is quadratic Eq. (2.19) or (2.20). In studying dynamic stability, it has been observed that dynamic relation for real power stability is mostly electromechanical while it is electromagnetic for the reactive power case.

To explore the steady state reactive power criteria, it is reasonable to assume the angles to be fixed. The voltages vary with reactive power drawn from the bus 2 in the manner shown by the reactive power (Q_2) vs. voltage difference ($V_1 - V_2$) characteristics in Figure 2.4.

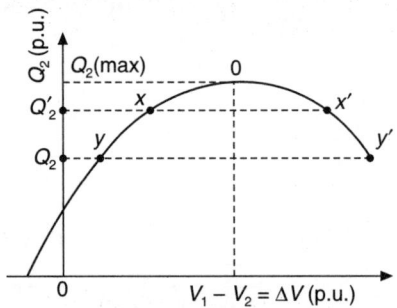

Figure 2.4 Q-ΔV characteristic to explain steady state stability of reactive power.

It may be observed that the basic character of the reactive power vs. ($V_1 - V_2$) curve for positive ($V_1 - V_2$) is almost same as the positive part of the P-δ curve as shown in Figure 2.3. Hence it can be commented that the system stability from the view point of reactive power status is maintained till Q_2 reaches Q_{max}. Beyond that position, the system's reactive power stability is lost and the system voltage at the receiving end will collapse even if the voltage at the sending end is maintained.

It may also be observed that no solution of the power flow equation will be obtained beyond point '0' indicating the collapse of system stability if the reactive power demand at load bus exceeds the limit given by Q_{2max}. Analysis shows that for any reactive load demand less then Q_{2max}, there are two operating points (i.e. two solutions). The solutions corresponding to operating point (x, y) etc. (below 0') are stable while points (x', y'), etc. (beyond 0') are not stable from the point of view of operational requirement. Q_{2max} is conventionally termed static reactive power stability limit in exactly the same way as P_{2max} has been treated[†]. However,

---

[†]
$$Q = f(V, \delta)$$
Therefore, for small perturbations,
$$\Delta Q = \frac{\partial Q}{\partial \delta} \Delta \delta + \frac{\partial Q}{\partial V} \Delta V$$

(Contd.)

$P_{2\max}$ lies on the curve which is a sine wave characteristic while $Q_{2\max}$ lies on a parabola.

Considering again the two stability limits ($P_{2\max}$ and $Q_{2\max}$), an arbitrarily small neighbourhood may be located very close to these two unique points. Let these areas be defined as

$$|\Delta P| = |P_{2\max} - P_2| = \varepsilon_1 \tag{2.21}$$

and
$$|\Delta Q| = |Q_{2\max} - Q_2| = \varepsilon_2 \tag{2.22}$$

where ε_1 and ε_2 are the errors, all greater than zero. Till this criterion of $\varepsilon > 0$ is satisfied, there will be two solutions of power flow equations located very close to each other. Theoretically ε_1 or ε_2 may be very small indicating very close pair of solutions of the load flow equations. The solution will tend towards a unique solution if ε_1 or $\varepsilon_2 \to 0$. This point (i.e. the steady state real or reactive power limit) is also termed 'bifurcation point'. The static stability problem of the power system is thus equivalent of finding the bifurcation point of the power flow equations to the transmission system.

2.5 REAL AND REACTIVE POWER TRANSIENT STABILITY

The transient stability is a concept unique to power systems and has already been introduced in textbooks and research publications. Transient stability is tied to specified disturbances (e.g. short-circuit, switching, loss of line) at specific places within the network and is governed by several countermeasures (e.g. clearing of short-circuit, etc.). An electrical transmission network possesses transient stability problems by virtue of contingencies. While encountering these problems, the transient real power stability is retained provided the system settles down to a stable operating state once the sudden disturbance is withdrawn. Its recovery will be total if voltages at its terminals recover to the near nominal values following the sudden disturbance and its withdrawal. Thus it is evident that similar to the steady state stability, transient stability also has two-fold appearance—the real power transient stability and the reactive power transient stability.

Continued

Applying the concept of decoupled load flow, $\dfrac{\partial Q}{\partial \delta} = 0$

Hence,
$$\Delta Q = \frac{\partial Q}{\partial V} \Delta V = C' \Delta V$$

and thus
$$\Delta V = \frac{\Delta Q}{C'}$$

Furthermore, ΔV will be positive till $C' > 0$, i.e. $\delta Q/\delta V$ is positive; ΔV will be negative as soon as $C' < 0$. Hence the system voltage will remain in stable position till $C' = 0$. Critical state is arrived at $C = 0$ and the system voltage will land in an unstable state as $C < 0$.

2.5.1 Real Power Transient Stability

The simplest two-area system is again assumed where the system losses are neglected. The shunt branches have not been taken into account for simplicity, and the system has been assumed to be in stable operation from the point of view of the real and reactive power flow.

Considering the specific operating load level at P_2' (Figure 2.5), let a specific amount of generation ΔP_2 be lost at the receiving side bus so that the load at that bus is raised to P_2'', and the system operation is shifted from the earlier stable operating point P_1 to another operating point P_2 corresponding to power demand P_2''. Obviously, this state of operation is not stable as the transmission line keeps the magnitude of power transfer as P_2' and not as P_2'' due to inertia of the system. The power angle remains the same at δ_2. As there is now more load in the receiving end bus than the power available at that bus, the generator supplying the load at this bus will start slowing down which results in the enhancement in the magnitude of $(\delta_1 - \delta_2)$ on δ_{12}'. This causes an increase in power output at bus 2 (receiving-end bus) along the sine curve.

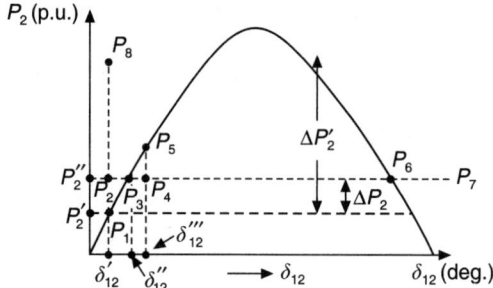

Figure 2.5 Transient operation of a power system on P–δ curve.

The power balance is restored at the state of operation of the system corresponding to point P_3 (angle δ_{12}''). But at this point of system operation, the machine in the receiving-end bus runs slower than that at the sending-end bus and hence $(\delta_1 - \delta_2)$ on δ_{12}'' increases further. The point of operation further shifts along the sine curve. However, if the sending-end bus can react in the mean time and transmit more power to meet the increase in load demand at the receiving-end bus, the machine at the receiving end may get the scope to speed up again. The new operating point may then be reached at point P_4 where P_4 indicates the load demand (P_2'') at new angle δ_{12}''' (initial operating angle was δ_{12}' corresponding to load demand P_2'). The power that can be transmitted being the power output corresponding to P_5 on the sine curve, the operation continues and δ_2 will go on decreasing resulting in the decrease in δ_{12} and the system may settle down, after a few damped oscillations, at a stable operating point corresponding to P_3. It is convenient to check the transient stability analytically applying the famous 'equal area criterion'. In this case, it can be conclusively commented that the real power transient stability is maintained provided the area $P_1P_2P_3$ equals the area $P_3P_4P_5$. At the equilibrium point P_3, the system readjusts the generation at the receiving end. A somewhat more insight into the problem reveals that the load demands P_2' and P_2'' may have some other values also and the swing of P_4 may be allowed till the operating point is designated by the operation at P_6. Equal area criterion being satisfied, transient stability will be maintained for these conditions too. Any movement of the system load demand beyond P_6 will make the system fully unstable.

Also, in case the initial operating point is at P_1, the load demand at the receiving-end bus being P_2', transient stability will be lost immediately if the loss of generation at the receiving-end is such that (in this case, let it be $\Delta P_2'$) the system apparent load demand moves to a point denoted by P_8. This time, the generator at the receiving-end bus will face continuous deceleration and the sending-end system will not have the capability to boost up power flow to meet this load demand. Thus $(\delta_1 - \delta_2)$ will go on increasing, pushing the system towards vulnerable collapse.

Thus it has been established that the transient stability is governed by the amount of sudden load change and the capability to cope with this sudden change in demand. It is also a function of the power angle between the buses and the instant when the sending end can transfer power to assist the receiving end.

2.5.2 Reactive Power Transient Stability

The reactive power vs voltage difference (Q vs $(V_1 - V_2)$) characteristic at the receiving-end bus is redrawn in Figure 2.6. Reactive power generation is Q_2'

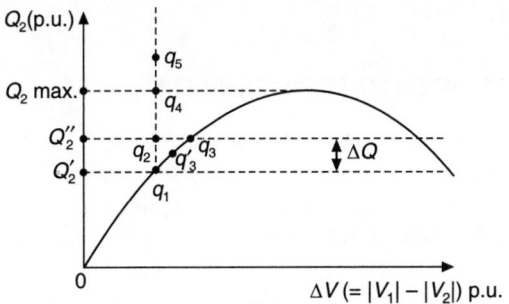

Figure 2.6 Transient operation of a power system on Q–ΔV curve.

when the system operates at q_1, with stable terminal voltages, and system frequency near or at nominal values. Let the reactive power supply at this bus drop due to failure of a capacitive bank which is identical to the sudden increase in reactive power at this bus. Let this increment be denoted by ΔQ. The operating point then shifts from q_1 to q_2. Due to inability of the transmission line to supply this reactive demand immediately, the bus voltage drops to E_2'. (This is analogous to the deceleration of the machine in Figure 2.5.) The new reactive power demand is

Q_2'' and the system operation may be stable at an equilibrium point q_3 on the reactive power characteristic corresponding to a lower system voltage. However, the alternator connected to the receiving-end bus will be over-excited to supply the additional reactive demand, and the system operation may be shifted to the stable operating point q_1 again. In case the transmission line is now able to transfer this additional reactive power from the sending-end bus, the result will be the same, i.e. the operating point will be shifted to q_1.

In case, the system does not possess the additional reserve of reactive power following the apparent increase in reactive power demand, the system may settle down to a new lower voltage state corresponding to the point q_3 on the curve. The transient stability is said to be retained in these cases in the event of sudden loss of reactive power supply to the receiving-end bus.

On the other hand, similar to the case of real power transient stability, if the magnitude of the loss of reactive power supply at the receiving-end bus is high enough to push the operating point beyond the point q_4, there will be no possible equilibrium point. The receiving-end bus bar voltage keeps falling and the system proceeds from point q_5 directly to voltage collapse. Under this condition, the system is said to have lost transient stability. [In case of transient stability problem of real power, the control of active power from the generators is carried out utilising governors while for the reactive power stability problem, the excitation is adjusted utilising an AVR. As AVR operation is faster than the governor operation, the service of reactive power reserve in the alternator may be brought into the receiving-end bus faster. Hence, the operating point may be at q_3' instead of q_3 while the AVR overexcites the alternator in the first case of transient stability problem considered, and the system operation may shift directly from point q_2 to q_3' and then to q_1.]

2.6 CONCEPT OF DYNAMIC STABILITY

Dynamic stability is associated with the normal operation following a minor disturbance. Hence, it is customary to study the dynamic stability in the small neighbourhood of one equilibrium, when the equilibrium point may theoretically be either in the stable or unstable zone of either of the real power vs. power angle or reactive power vs. voltage curve (Figure 2.3 or 2.4). Whether the system is considered in real or reactive power sense, it is said to be dynamically stable only when it, after any sufficiently small perturbation, returns to the equilibrium point. This condition is analogous to the convergence of load flow solution for incremental changes in power flow condition. However, without converging to the stable point, if the load flow solutions diverge, the system cannot reach the stable equilibrium point and then the system is said to be dynamically unstable.

In long lines, the return of the system, after encountering small perturbation, may be asymptotic giving rise to the fluctuation of power and voltage, commonly known as power and voltage oscillation. Under this condition, PSS (Power System

Stabiliser) or SVC (Static VAR Compensator) is frequently needed to ensure the damping of power and voltage oscillations.

2.7 RELATION BETWEEN VOLTAGE AND REACTIVE POWER AT A NODE IN A POWER SYSTEM

In this section, analytical expressions will be developed to illustrate the sensitivity of the voltage to the reactive power for any perturbation in the operating power status of the system.

Observation of the reactive power expression i.e. Eq. (1.98), reveals that the system voltage is a function of real and reactive power, i.e.

$$V = f(P, Q) \tag{2.23}$$

Therefore, for small perturbations,

$$dV = \frac{\delta V}{\delta P} dP + \frac{\delta V}{\delta Q} dQ$$

$$= \frac{dP}{\left(\frac{\delta P}{\delta V}\right)} + \frac{dQ}{\left(\frac{\delta Q}{\delta V}\right)} \tag{2.24}$$

It can be seen from Eq. (2.24) that, the change in voltage at the specific node is governed by the quantities $(\delta P / \delta V)$ and $(\delta Q / \delta V)$.

It has been shown in the preceding chapter that the voltage regulation in a transmission line of a radial system is composed of inphase and quadrature voltage drops Eq. (1.95), where the inphase voltage drop plays the principal role in determining the magnitude of net regulation. The effect of quadrature drop is negligible and with this reasoning, Eq. (1.94) becomes

$$\Delta V = \frac{R_l P + X_l Q}{V} \tag{2.25}$$

or

$$E - V = \frac{R_l P + X_l Q}{V}$$

or

$$(E - V)V - R_l P - X_l Q = 0 \tag{2.26}$$

Assuming a constant sending-end voltage, from Eq. (2.26),

$$\frac{\delta P}{\delta V} = \frac{E - 2V}{R_l} \tag{2.27}$$

and

$$\frac{\delta Q}{\delta V} = \frac{E - 2V}{X_l} \tag{2.28}$$

Substituting the values of $\left(\dfrac{\delta P}{\delta V}\right)$ and $\left(\dfrac{\delta Q}{\delta V}\right)$ from Eqs. (2.27) and (2.28) respectively in Eq. (2.24), we get

$$dV = \dfrac{dP}{\dfrac{E-2V}{R_l}} + \dfrac{dQ}{\dfrac{E-2V}{X_l}}$$

$$= \dfrac{dP \cdot R_l + dQ \cdot X_l}{E-2V} \qquad (2.29)$$

In a lossless system, $R_l = 0$. Also applying the concept of decoupling, the term $\delta P / \delta V$ is of greater interest. Hence, Eq. (2.29) may be rewritten as

$$dV = dQ \cdot \dfrac{X_l}{E-2V}$$

or

$$\dfrac{dQ}{dV} = \dfrac{E-2V}{X_l} \qquad (2.30)$$

It is thus evident that the smaller the reactance associated with the node, the larger is the value of (dQ/dV) for a given voltage drop. In other words, it means that for a low reactance system, the voltage drop is inherently small. In the power system, due to limited number of EHV lines and low system short-circuit strength, the reactance of the system is high. This causes a higher voltage drop in the system. However, this drop can be compensated by artificial injection of reactive power, thus making dQ/dV higher. It enhances the cost of the system.

Also, in case of light loading, the receiving-end voltage is approximately the same as the sending-end voltage and thus Eq. (2.30), nominalising the reactance x_l as x, becomes

$$\dfrac{dQ}{dV} = -\dfrac{E}{X} \qquad (2.31)$$

But E/X represents the short-circuit current flowing through the system. Hence it is seen that the magnitude of (dQ/dV) is equal to the magnitude of *short-circuit current*. Systems with higher system reactance indicate lower magnitude of short-circuit current (i.e. lower short-circuit strength) and this has also been indicated earlier.

2.8 REACTIVE POWER REQUIREMENT FOR CONTROL OF VOLTAGE IN LONG LINES

In the earlier discussions, it has been established that voltage control is a problem in long lines and the problem is further aggravated due to high terminal reactances. It is necessary, at this stage, to establish the reactive power requirement of long lines in order to maintain the system voltage near the rated value (i.e. when the receiving-end voltage varies from 0.95 p.u. to 1.05 p.u., the rated value being 1.0 p.u.)

It is advantageous to use the generalised line equation for the line including the source reactance and the terminal transformers.

Thus,

$$E = AV + BI_R \tag{2.32}$$

Assuming the phase angle as ϕ_R at the receiving end (i.e. the load current lags behind the receiving-end voltage by angle ϕ_R), and taking V as the reference phasor,

$$A = A \angle \alpha, \quad B = B \angle \beta \tag{2.33}$$

Thus from Eq. (2.32),

$$E \angle \delta = AV \angle \alpha + BI_R \angle (\beta - \phi_R)$$

or $E \angle \delta = AV \cos \alpha + jAV \sin \alpha + BI_R \cos(\beta - \phi_R) + jBI_R \sin(\beta - \phi_R)$

Equating the magnitudes only,

$$E^2 = A^2 V^2 + B^2 I_R^2 + 2ABV I_R [\cos(\alpha - \beta) \cos \phi_R - \sin(\alpha - \beta) \sin \phi_R] \tag{2.34}$$

But
$$\cos \phi_R = \frac{P}{VI_R} \quad \text{and} \quad \sin \phi_R = \frac{Q}{VI_R} \tag{2.35}$$

for the loading of $(P + jQ)$ at the receiving-end bus.

Substituting the values of $\cos \phi_R$ and $\sin \phi_R$ from Eq. (2.35) in Eq. (2.34),

$$E^2 = A^2 V^2 + B^2 I_R^2 + 2ABP \cos(\alpha - \beta) - 2ABQ \sin(\alpha - \beta)$$

$$= A^2 V^2 + B^2 (I_{PR}^2 + I_{QR}^2) + 2ABP \cos(\alpha - \beta) - 2ABQ \sin(\alpha - \beta)$$

[where, I_{PR} = real part of I_R and I_{QR} = reactive part of I_R]

$$\therefore \quad E^2 = A^2 V^2 + B^2 \left[\frac{P^2}{V^2} + \frac{Q^2}{V^2} \right] + 2ABP \cos(\alpha - \beta) - 2ABQ \sin(\alpha - \beta) \tag{2.36}$$

Hence for a given system, A and B being known as generalised circuit constants, Q may be calculated for any specific amount of real power flow to make the receiving-end voltage of desired value in terms of the sending-end voltage.

2.9 OPERATIONAL ASPECTS IN REACTIVE POWER AND VOLTAGE CONTROL

In power systems, EHV networks have been constructed where the transmission lines of these networks often link the remote hydro, thermal and nuclear power plants with the load centres, transporting the large blocks of electric power to load centres from these power stations. Interconnected with this EHV system are the numerous generators, transformers, reactors, capacitors, etc. that are directly or indirectly needed for rendering uninterrupted electricity supply to the consumers at desired voltage level and to various locations.

Each series element in this system has a reactance which has a reactive power loss proportional to the square of the current passing through it. Reactive power losses are a system-wide phenomenon and can escalate during heavy loading and may diminish during light loading. Any series winding having an inductance, serves as a sink for reactive power while the capacitances of EHV lines and cables act to charge the reactive power in the system. Enhancements in the system voltage level invite the reactive power mismatch problem since EHV lines have high *(X/R)* ratio that increases the series loss due to heavily loaded lines and transformers, causing more reactive power to be drained. During light load periods, the reactive power surplus due to high charging and low series reactive loss may affect the system operation by causing overvoltage in the system (Ferranti effect).

Much as frequency is a measure of the balance between the generation and the real load throughout a power system, transmission voltage levels indicate the balance between the supply and demand of reactive power. Although under a specific operating condition, frequency is uniform throughout the power system, voltage levels can vary at different points of the transmission network due to the reactive power problem.

Voltage control problem, in its extreme stage, may cause voltage instability which is basically a reactive power mismatch in a power system as shown in the preceding discussion. This may occur due to steady, transient and dynamic state variations in system operating conditions (change in loading, transformer tap position, alteration of alternator outputs, switching of capacitors, reactors, outages, etc.).

2.9.1 System MVAR Mismatch

Two possibilities exist for the occurrence of the reactive power mismatch, namely (1) reactive power surplus and (2) reactive power deficit. The disturbance that ultimately causes these phenomena would most probably start within one region and dissipate the regional reactive power reserve, and then crawl to affect the remainder of the network. The loss of co-ordinated reactive power control, both for a surplus and a deficit of reactive power, is indicated by a significant change in:

    (a) transmission voltage at major substations and,
    (b) summated reactive power output of nearby rotating machines.

Reactive power surplus

The danger posed by a system reactive power surplus, is of overvoltage, which if excessive and sustained can result in insulation breakdown. To diminish the duration of these overvoltages, there is an essential need for fast countermeasures, and this can most effectively be achieved by tackling the overvoltage at each of the affected locations.

Countermeasures: Many shunt reactors are located at numerous substations and also at subtransmission levels to provide an effective means of handling a reactive power surplus. Adjustments of excitation of rotating generators is also a viable alternative.

Reactive power deficit

The sharp regional voltage reduction draws on inherent stabilizing factors which initially counteract a reactive power deficit. This voltage reduction reduces the nearby load magnitudes and simultaneously pushes the adjacent rotating units into the field, forcing more to draw the untapped reactive power reserve. In case the system is equipped with additional reactive power sources, the system voltage can settle down to a lower magnitude. Otherwise the system voltage stability may be lost.

Countermeasures: There are some methods by which this problem can be counteracted. These are as follows:

(a) Generator terminal voltage increase,

(b) Generator transformer tap changing,

(c) Reactive power boost,

(d) Quick acting load transformer tap changing,

(e) Strategic load shedding.

All these actions/operations can be done sequentially or separately depending upon the severity of deficit and effectiveness of the scheme. However, beyond a specified limit, transformer tap changing does not serve the purpose effectively and may push the system towards instability without assisting it.

2.9.2 Vulnerable System Disturbance

There have been a lot of system failures that occurred in the past in several countries due to the reactive power mismatch. Some of the critical system failures have been the North American Grid Disturbance in 1965, New York collapse in 1977, France National grid collapse 1978, Edison Company power failure in 1977, Belgian blackout in 1982, Northern grid disturbance in India in the years1984, 1987 and 1994, Eastern grid collapse in India in the years 1989, 1991, and 1997, 2000, 2004 and even later. These events got initiated from sudden small disturbances in EHV or HV grids, leading to reactive mismatch. These soon went beyond control by successive trippings and inter-trippings and ultimately fragmented the power system into sections, thus destroying the complete reliability of the affected power system. The probable reasons of these uncontrollable phenomena may be accounted as under:

(a) Their occurrence mostly could not be predicted.

(b) Except for the variations in voltage, there was no other way to detect these disturbances from the central control centre at the early state of disturbance.
(c) The spread of disturbances was so fast that it left no time for the human operator to rectify the basic problems.
(d) During the cascade trippings and when the power system reached at the verge of collapse, the coordination between numerous control centres became ineffective and uncertainty among the operators led to further failure of the system.

Studying the past cases of the breakdowns, it has been observed that the disturbance usually starts from the distribution system and/or from the transmission system in most of the cases, and the generating system has been very realy the source of such disturbances.

2.10 BASIC PRINCIPLE OF SYSTEM VOLTAGE CONTROL

In the operation of a power system, there is a continuous control of voltage levels. For sub-transmission and distribution, the voltages are controlled by tap-changing transformers. However, for the core of the network, i.e. in the transmission system, the voltage levels are maintained by drawing of the reactive reserves of the systems' controllable plant which is made up mainly of rotating units.

For major system reactive power disturbances, larger demands are met with the reactive reserves, and with automatic control providing the short-term correction of voltage levels.

2.11 REACTIVE POWER FLOW CONSTRAINTS AND THEIR IMPLICATIONS IN LOSS OF VOLTAGE

In this section, implications of reactive power mismatch in following a contingency in the transmission network (or at the load centre) have been established in the lossless frame. The advantages of using adequate effective reactive reserve at the load centre along with the utility of high terminal short-circuit strengths at the source and load nodes have also been highlighted.

State prior to disturbance

The system is presumed to remain under medium to heavy loading and some of the key EHV lines are fully loaded or might be to some extent overloaded due to adjacent EHV line(s) being out of service. However, the system frequency is within the tolerable limits and almost normal operating conditions prevail.

Case I: Initiating Event of Disturbance in Transmission System

The initiating disturbing event may be the loss of a highly loaded EHV transmission line, due to operation of its protective relays situated between a generation source and a major load centre. This tripping will cause an immediate extra loading of the adjacent EHV line(s) causing as substantially increased reactive burden on the system.

Immediate post disturbance situation: Immediately after the EHV line has been tripped, there would be a considerable voltage reduction at the adjacent load centres, thereby causing a significant load reduction. This in turn causes less demand in MW on the generating station and the system frequency may rise marginally till regularised by governor control. In the meantime, due to increased transmission line reactive loading, the automatic voltage regulator of the alternators would operate to restore the generator terminal voltage.

In case the system does not have enough reactive power reserve, a voltage reduction, following the disturbance, will be observed at all segments of the network down to the loads at distribution voltage levels. It also decreases the charging capacity of cables, lines and shunt capacitors throughout the subtransmission and distribution networks. In case the tap-changing operation has started, the distribution voltage is slowly restore to normal level. But each tap change will increase the voltage of the secondary (i.e. the voltage at LV winding) and concurrently increase the load MW and MVAR supplied by distribution transformers. This extra load, percolating through the subtransmission network, will cause its voltage to fall as each tap change raises the distribution voltage. In the subtransmission network, this will increase the series reactive power losses and also reduce line, cable and capacitor charging. If the EHV line is heavily loaded, the extra reactive loss for each extra increment of loading becomes quite substantial and is even more pronounced if the EHV level reduces. This is evident from the fact that when a transmission line surpasses SIL (Surge Impedance Loading), there is a significant rising rate of reactive losses. Further, the value of SIL reduces with the square of the voltage level as voltage falls. In extreme cases, there may be several MVAR of reactive line loss for each extra MVAR received at the load centre.

Should the tap-changing process continue unchecked on a vulnerable network, distribution voltage will recover fully, subtransmission voltage may recover partly, but the transmission voltage will continue to fall. Even though the load connected to the load centre may not increase physically when the voltage at the load centre is restored by the transformer tap-changing process, the transmission system requires extra reactive power which could be a final blow to system stability.

Each load centre transformer tap-change would directly increase the loads which would need to be supplied by the system generators. The reactive load increase on the generator would be enhanced by the additional series losses in the line and transformer. While the automatic voltage regulators maintain the generator terminal voltage, the extra reactive loading through the 'Gen-Transformer' would reduce the power station EHV level compounding with the series loss. With inadequate reactive reserves, system generators would attain and then exceed their continuous rated field currents. The result is the severe voltage crisis within the system.

CASE II: INITIATING EVENT OF DISTURBANCE IN DISTRIBUTION SYSTEM (CONTINGENCY IN THE DISTRIBUTION SYSTEM)

Assuming the operation of a protective gear, say in the static reactive power injection device (i.e. the static capacitor bank) in the load centre, the sequence of events can be analysed to predict the voltage control problems at the load centre. With tripping of the capacitor bank, load bus voltage drops are compounded by reactive power shortage due to fall in the charging capacity of lines and cables. Real demand may also fall due to lower residential loads with drop in voltage.

With On Load Tap Changer (OLTC) in the distribution centre in operation, the subtransmission voltage and thus the distribution voltage attain near normal levels while the main load bus voltage remains depressed. As a result the real demand on the load bus increases and may slightly overshoot the precontingency level. However, as the active power on the line increases, the voltage drop required to deliver the deficit reactive power also increases, and is compounded by an increase in series reactive losses and fall in reactive charging. The situation further deteriorates if the reactive power limits of the nearby generators are exhausted. With OLTC operation uninterrupted, the distribution voltage is retained near the nominal value by the transformer, while the voltage in transmission and subtransmission zones falls steadily. In a reactive power constrained line, this situation may lead to the voltage collapse at the load end depriving the system from having a stable operating point, even if it did not have any frequency instability in the precontingency period. During the steady fall of the load voltage, the collapse may accelerate due to overloading of the motors in the system. However, at the verge of collapse, there will be a transient voltage overshoot due to release of motor contactors and extinction of fluorescent lamps. This overshoot may again result, in an enhancement in real power demand, thus accelerating the voltaic collapse.

2.12 EFFECT OF TRANSFORMER TAP CHANGING IN THE POST-DISTURBANCE PERIOD

It is evident from the preceding discussion that if the subtransmission and distribution transformer tap changers function before the full restoration of subtransmission voltage by the EHV substation transformer, there will be an overshoot of EHV substation real and reactive loading.

Hence, with reference to transformer tap changing, the following factors should be carefully considered.

1. Timing of transformer tap changing should be graded such that the higher the voltage the faster is the tap changing.
2. Load overshoot of both MW and MVAR can be avoided if the EHV substation transformer tap changer can restore the subtransmission voltage levels before any downside transformer tap changer functions.
3. Normal tap-changing will restore the system load to predisturbance levels, even if EHV level is falling. Under this condition, strategic load shedding may be effective.

4. System load overshoot due to transformer tap-changing action may affect the balance unfavourably to system voltage stability in an emergency situation.

2.13 EFFECT OF GENERATOR EXCITATION ADJUSTMENT IN THE POST-DISTURBANCE PERIOD

A sharp, sustained reduction in transmission voltage levels would cause field forcing and over-excitation of nearby rotating units. This field forcing will provide access to a normally untapped level of reactive power reserve. If it can be sustained, this over-excitation will expand the power system's range of reactive power co-ordination and hence its voltage stability.

In case, the over-excitation is not permissible and when there is no effective reactive reserve, transmission voltage will fall putting the system on risk of system collapse.

In an extensive power system, this effect would mushroom outwards, so that progressively more and more units would soon become over-excited and would then have their over-excitation restricted. At the epicentre, voltage would fall to remarkably low levels. It has been recovered from the records that during the 1978 France disturbance, Paris power stations were islanded due to the above condition when the voltage dropped below 0.7 p.u.

2.14 PRACTICAL ASPECTS OF REACTIVE POWER FLOW PROBLEMS LEADING TO VOLTAGE COLLAPSE IN EHV LINES

Voltage control problem is a typical problem encountered in dealing with the operation of transmission lines. This problem typically falls into one of the following categories:

(a) *Long transmission lines:* *In* power systems, long lines with voltage uncontrolled buses at the receiving ends create major voltage problems during light load or heavy load conditions. When the loading level is low, i.e. below SIL, the receiving-end voltage rises while for heavy loading conditions (much above SIL) the receiving end voltage drops (this is evident from the reactive power characteristics of a line). The receiving-end voltage case of heavy loading conditions, the series reactive drop may be very high causing voltage depression at the receiving end. In extreme cases, the reactive power demand (i.e. series reactive loss) of the line may be enormously high, causing severe voltage control problems. This voltage depression cumulatively increases due to the lower charging capacity and the higher reactive current intake in the induction motor loads. Any contingency further deteriorates the condition.

(b) *Radial transmission lines:* In a power system, most of the parallel EHV networks are composed of radial transmission lines. Any loss of an EHV line in the network causes an enhancement in system reactance. In case a condition appears where the increase in reactive power delivered by the lines(s) to the load for a given drop in voltage, is less than the increase in reactive power required by the load for the same voltage drop, a small increase in load puts the system in an unstable state. On load tap changing, transformer improves or maintains the distribution *voltage* while there is no improvement in transmission voltage. This also affects the voltage stability and has been discussed earlier.

(c) *Shortage of local reactive power:* It has been explained earlier that there should be a perfect match between reactive power generation and demand at any bus to have voltage stability at that bus. However, there may be a disorganised combination of outage and maintenance schedule that may cause localised reactive power shortage and lead to voltage control problems. In a reactive power constrained system, the reactive power reserve in the network is low. A disturbance in a load bus may cause a shortage of reactive power supply and/or an increase in reactive power demand. The result is a voltage drop in the transmission line. This gives rise to the problem of voltage control and also reduces the charging capacity of reactive power of the system. Until there is a local reactive power support, any attempt to import reactive power through the long EHV lines is not successful and does not serve the purpose. Under this condition, the bulk system voltage drops down.

(d) *Reactive dispatch capability:* This problem is most likely to occur in any system where a common trend is to use local generators to supply reactive loads and losses. At normal operating voltage the reactive power availability from generators, capacitors, and line and cable charging balance the reactive load as well as the reactive losses in line and terminal transformers. In case of a drop in system voltage, line loading being constant, the reactive loss increases and the charging capability of the system decreases. The result is an increase in the reactive power output of the generator. Thus, the voltage drop initiates an increase in reactive power output of the local generators. In case of a continuous increase in reactive power demand at the load bus or due to voltage drop in that bus, the, reactive power output of the generators goes on increasing. However, this enhancement is limited due to thermal limits of the generator(s). Any attempt to change the generator voltage set-point by alteration in the AVR system will cause enhancement of generator voltage. This may further enhance loading on the bus or cause enhancement of reactive flow from it to other generators creating undesired operation. In a weak capacity power system, the operator of such a generator bus may notice the rise in the reactive output of the alternators as the problem of undervoltage is attempted to be solved. The other alternative is to bring the reactive power from remote sources. However, this is not a very wise solution as discussed earlier. The net result is the undesired drop in the receiving-end voltage.

(e) *High voltage problem:* This is another problem to be tackled in any transmission system. This has already been discussed and has conventionally been termed Ferranti effect.

EXERCISES

1. What is decoupling concept? Show that for a lossless transmission two-bus model, Q-V and P-δ quantities are closely coupled.
2. How would you prove that real power loading has got implication in reactive power demand?
3. What is the role of reactive power demand on the voltage regulation of a power line?
4. Show that for high values of power angle, (dQ/dV) approaches zero value and indicates loss of stability.
5. Briefly explain the meaning of 'load bus reactive power sensitivity'.
6. What do you mean by static real stability and reactive power stability.
7. How would you explain the real power transient stability problem for a sudden increase in load demand?
8. Explain the transient power reactive stability problem in a two-bus power line.
9. Develop a relation between voltage and reactive power at a node in a power system.
10. Develop an expression to find the magnitude of reactive power requirement for voltage control in long lines.
11. How will you explain the implication of reactive power mismatch in system operation?
12. What is reactive power flow constraint and what is its role in loss of system voltage?
13. Explain the effect of tap-changing transformer and automatic voltage regulator of generator in voltage control of a transmission system.
14. List a few practical aspects for describing the reactive power flow problem in voltage collapse.

Chapter 3

Voltage Stability—Fundamental Concepts

3.1 INTRODUCTION

Voltage stability is a problem in power systems which are heavily loaded, faulted or have a shortage of reactive power. The nature of voltage stability can be analysed by examining the production, transmission and consumption of reactive power. The problem of voltage stability concerns the whole power system, although it usually has a large involvement in one critical area of the power system. Maintaining synchronism between the various parts of a power system becomes increasingly difficult, as the systems and interconnections between the systems continue to grow.

This chapter describes the voltage stability phenomena. First voltage stability, voltage instability and voltage collapse are defined and the aspects of voltage stability are classified. With a short example of maximum transfer capacity being described, the modelling and the effect of power system components on the long-term voltage stability studies are highlighted.

3.2 REACTIVE POWER AND VOLTAGE COLLAPSE

Voltage collapse typically occurs in power systems which are usually heavily loaded, faulted and/or have reactive power shortages. Voltage collapse is a system instability and it involves many power system components and their variables at once. Indeed, voltage collapse often involves an entire power system although it usually has a relative larger involvement in one particular section of the power system.

Though a number of variables are typically involved, some insight into the physical nature of voltage collapse may be gained by examining the generation, transmission and consumption (including surplus and deficit) of reactive power. Limitations on the production of power include generator and reactive power compensators' limits and the reduced capacitive reactive power by the line and fixed capacitors due to low voltages. The primary limitations on the transmission of power are the high reactive losses on heavily loaded long transmission lines. There may also be outages and reduced transmission capacity. Reactive power demands of loads increase with load increase, motor stalling or changes in load composition (when the reductive part of the load current increases).

Voltage collapses can occur in a transient time-scale or in a long-term time-scale. Voltage collapse in the long-term time-scale can include effects from the transient time-scale; for example, a slow voltage collapse taking several minutes may end in a fast voltage collapse in the transient time-scale.

3.3 CHANGES IN POWER SYSTEM CONTRIBUTING TO VOLTAGE COLLAPSE

There are several power system disturbances which contribute to voltage collapse:

(a) Increase in inductive loading
(b) Reactive power limits attained by reactive power compensators and generators
(c) OLTC operation
(d) Load recovery dynamics
(e) Generator outage
(f) Line tripping.

Most of these factors have significant effects on reactive power production, transmission and consumption. Switching of shunt capacitors, blocking of OLTC operation, generation rescheduling, bus voltage control, strategic load shedding and allowing temporary reactive power overloading of generators may be used as some of the effective countermeasures against voltage collapse.

3.4 CONCEPT OF STABILITY OF TRANSMISSION SYSTEM

A very common and widely accepted definition of "Stability" refers to small disturbance stability of an operating point and is defined as follows:

An operating point of a power system is "small disturbance stable" if, following any small disturbance, the power system state returns to the identical or close to the predisturbance operating point. Thus a power system at a given operating point is voltage stable if the load voltage approaches the post-disturbance steady value following a given disturbance. Traditionally the stability problem had been the rotor angle stability, i.e. maintaining synchronous operation. Instability may also occur without loss of synchronism and in that case the concern is the stability of voltage. The small disturbance voltage stability corresponds to a linearised dynamic model with eigenvalues having negative real parts.

In the power system, if the change is gradual, such as for the case of slow load increase, the restabilisation causes the power system to approach the stable operating point as this point gradually changes. This operation of the power system is usual and normal. However if the stable operating point disappears due to some disturbance, the power system can loss stability following a change. The dynamic fall of voltage in this case is identified as voltage collapse. Thus, voltage instability is the absence of voltage stability and results in progressive voltage decrease. Rapid increase in load or large power transfer may be instrumental in causing voltage instability. The instability is almost always an *aperiodic* decrease in voltage. Oscillatory voltage instability may also be possible.

The voltage collapse following any sudden disturbance may be treated as treatment process and this may also result in dynamic fall of voltage. The collapse can be complex, with an initial slow decline of voltage. With further change in the system, a fast decline in voltage may take place. Thus a transient collapse can include dynamics at either or both of long-term and transient time-scales. Proper corrective action may save the system from voltage collapse.

Voltage stability can also be called load stability. A power system does not have the capability to transfer an infinite amount of electrical power to load. The main cause for voltage instability is the inability of power system to meet the demands for reactive power in heavily stressed systems.

3.5 RELATION OF VOLTAGE STABILITY TO ROTOR ANGLE STABILITY

Voltage stability and rotor angle (or synchronous) stability are more or less interlinked. Slower forms of voltage stability and small disturbance rotor angle stability are interlinked and so also are the transient instability of voltage and rotor angle. Often, the mechanisms are difficult to separate.

Voltage stability is concerned with load areas and load characteristics. Rotor angle stability is basically the generator stability while the voltage stability involves the loads and the means for voltage control. Thus, sometimes it is called the load stability. Both the forms of stability require reactive power support, though in large interconnected systems voltage collapse of load area is possible without the loss of synchronism of any generator.

3.6 STABILITY MARGINS

When a system changes from one operating condition to another, the most important question that immediately strikes us is 'how stable is the new operating condition?' The question indirectly asks about the margin of stability in the new position.

Any margin of stability determines the maximum change in the related parameter for which the margin of stability is being ascertained, when the operating point of a system shifts from one position to a new position.

Power system stability is directly associated with the stability of its performance parameters like real, reactive and dynamic stabilities, both in steady state and during the transient state of operations. Another important stability parameter is the knowledge about voltage stability which is also similarly associated with system stability utilising the concept of stability margin. The real power steady state stability margin is defined as

$$K_{P_{st}} = \frac{P_{max} - P_0}{P_0} \times 100 \text{ in } \% \tag{3.1}$$

which in other form also can be measured (Figure 3.1) as

$$K_{\delta_{st}} = \frac{\delta_{max} - \delta_0}{\delta_0} \times 100 \text{ in } \% \tag{3.2}$$

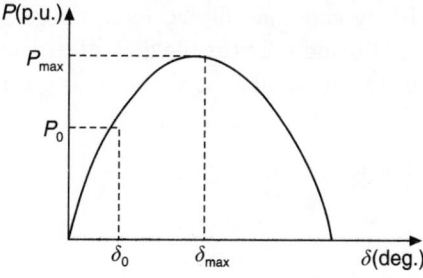

Figure 3.1 *P–δ* curve to explain real power steady state stability margin.

[It may be noted that δ_{max} has been assumed to be 90° in Figure 3.1 which in real cases may be above or below 90° depending on the level of excitation of the machine (Figure 3.2). Hence, the magnitude of δ_{max} depends on the system operation and does not have any absolute value. In cases of reactive power constrained lines, *it has been shown earlier that the limiting value of the power angle in a two area simple radial type power system is given by* $(\pi/4 - \phi/2)$ *which*, for unity power factor load may be 45° only. In that case, the steady state stability margin, being governed by the voltage stability and load bus power factor, is far lower than that shown in Figure 3.1 or Figure 3.2]

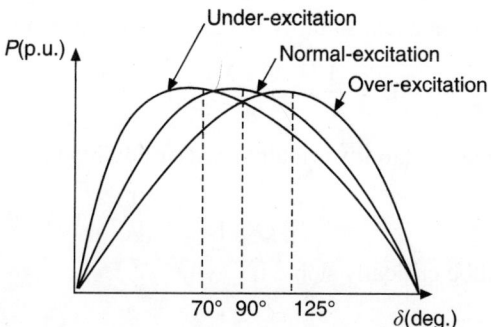

Figure 3.2 Role of excitation of steady state stability.

With similar reasoning the steady state reactive power stability margin is given by

$$K_{\theta_{st}} = \frac{Q_{2\max} - Q_2}{Q_2} \times 100 \text{ in } \% \qquad (3.3)$$

The transient stability criterion is represented in a slightly different manner. Referring to Figure 3.3, if A_1 and A_2 be the areas of acceleration and deceleration, assuming the loss of one EHV line between two generators, the system reactance increases causing drop in real power flow. Once the real power flow is dropped, the mechanical power input being the same, the machine will accelerate till the power output again increases. But due to inertia, the rotor continues to move past the equilibrium point and then starts decelerating as the electrical power is more than the mechanical power. However, the decelerating mode may not stop at the equilibrium point and may continue till acceleration starts again. Following a transient disturbance, ΔA being the possible area of deceleration till the power angle reaches $\delta_{\max}$, the area of possible deceleration is given by

$$\Delta A = A - A_1 \qquad (3.4)$$

where A = total area of deceleration.

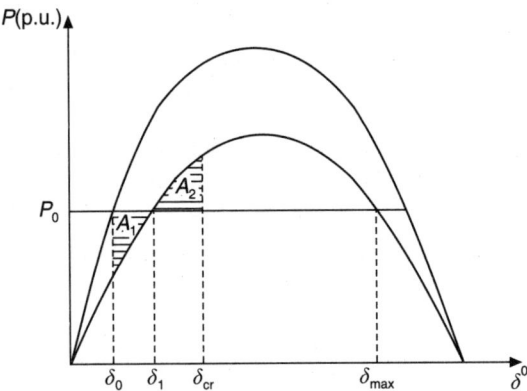

Figure 3.3 Equal area criterion.

The criterion of transient stability in real power is given by

$$C = \frac{A}{A_1} = \frac{A_1 + \Delta A}{A_1} = 1 + \frac{\Delta A}{A_1} \qquad (3.5)$$

Obviously, the transient stability is maintained (following the equal area criterion) if

$$C > 1$$

The condition will be critically stable if

$$C = 1$$

and instability occurs when $C < 1$.

The transient stability margin $(K_{P_{tr}})$ real power is thus given by

$$K_{P_{tr}} = \frac{A - A_1}{A_1} \times 100 \text{ in \%} \qquad (3.6)$$

Also,

$$K_{\delta_{tr}} = \frac{(\delta_{max} - \delta_1) - (\delta_1 - \delta_0)}{(\delta_{cr} - \delta_0)} \times 100$$

$$= \frac{\delta_{max} + \delta_0}{\delta_{cr} - \delta_0} \times 100 \text{ in \%} \qquad (3.7)$$

if the transient stability margin of real power is expressed in terms of angles.

Dynamic stability is always associated with small perturbations and hence in a stable system, it is frequently measured in terms of margin whose concept is identical to that of steady state margins. Both these terms are frequently termed in an analogous manner in order to describe the real and reactive power stability margins.

Voltage stability, being one of the most important parameters in the present day stability studies, is frequently associated with its measurement. The assessment of margin of voltage stability is very important in determining the capability of the system to cope with the problems that cause voltage instability.

Most of the studies so far reported deal with the steady state voltage instability problems. The process of voltage collapse is usually slow since with deteriorating voltage at the load bus, the motor loads are thrown off leading to enhanced load bus voltage. Moreover, under the voltage crisis periods, it is a conventional trend to boost up the distribution voltage by transformer tap change even though the transmission voltage remains in crisis. This may increase the bus loads again by reconnecting the industrial loads. This effect results in the deterioration of reactive power position and the series reactive loss may increase tremendously, depressing the load bus voltage to an alarming level. In the process of declining system voltage, the local generators will also try to elevate their reactive power output till their thermal limit of rotor winding current is reached. The remote generators will also try to press more reactive power through the line which will also ineffectively try to boost up the system voltage.

At the fag end of the entire process, the system voltage may collapse even if any effective means to improve voltage stability were adopted. However, the above steps of the process clearly indicate that the mechanism of voltage collapse is a slow one. Thus it is customary to represent the voltage instability mostly as a steady state problem.

The voltage stability margin, at steady state, is given by

$$K_{v_{st}} = \frac{V_0 - V_{cr}}{V_0} \times 100 \text{ in p.u.} \qquad (3.8)$$

However, due to sudden loss of reactive power compensators in the load bus, there may be transient voltage instability, particularly where the load bus voltage is governed by the load stability. Also, due to voltage swings, transient voltage instability may develop at the load bus when it is fed from two radial lines. Hence, it is a present day development to define the term transient voltage stability margin as:

$$K_{V_{tr}} = \int_0^t \left[\frac{V(t) - V_0}{V_0} \right]^2 dt \tag{3.9}$$

3.7 DEFINITION AND CLASSIFICATION OF VOLTAGE STABILITY

3.7.1 Definition of Voltage Stability, Voltage Instability and Voltage Collapse

Power system stability is defined as a characteristic for a power system to remain in a state of equilibrium at normal operating conditions and to restore an acceptable state of equilibrium after a disturbance. Traditionally, the stability problem has been the rotor angle stability, i.e. maintaining synchronous operation. Instability may also occur without loss of synchronism, in which case the concern is the control and stability of voltage. Voltage stability is defined as follows: "The voltage stability is the ability of a power system to maintain steady acceptable voltage at all buses in the system at normal operating conditions and after being subjected to a disturbance."

Power system is voltage stable if voltages after a disturbance are close to voltages at normal operating condition. A power system becomes unstable when voltages uncontrollably decrease due to outage of equipment (generator, line, transformer, bus bar, etc.), increment of load, decrement of production and/or weakening of voltage control. Another definition of voltage instability is "Voltage instability stems from the attempt of load dynamics to restore power consumption beyond the capability of the combined transmission and generation system." Voltage control and instability are local problems. However, the consequences of voltage instability may have a widespread impact. Voltage collapse is the catastrophic result of a sequence of events leading to a low-voltage profile suddenly in a major part of the power system.

Voltage stability can also be called "load stability." The main factor causing voltage instability is the inability of the power system to meet the demands for reactive power in the heavily stressed system to keep desired voltages. Other factors contributing to voltage stability are the generator reactive power limits, the load characteristics, the characteristics of the reactive power compensation devices and the action of the voltage control devices. The reactive characteristics of AC transmission lines, transformers and loads restrict the maximum of power system transfers. The power system lacks the capability to transfer power over long distances or through high reactance due to the requirement of a large amount of reactive power at some critical value of power or distance. Transfer of reactive power is difficult due to extremely high reactive power losses, which is why the reactive power required for voltage control is produced and consumed at the control area.

3.7.2 Classification of Power System Stability

Power system stability is classified above as rotor angle and voltage stability. A classification of power system stability based on time-scale and driving force criteria is presented in Table 3.1. The driving forces for an instability mechanism are named generator-driven and load-driven. It should be noted that these terms do not exclude the effects of other components on the mechanism. The time-scale is divided into short and long-term time-scales.

Table 3.1 Classification of power system stability

Time-scale	Generator-driven	Load-driven
Short-term	Rotor angle stability Small-Signal, Transient	Short-term voltage stability
Long-term	Frequency stability	Long-term voltage stability Small disturbance, Large disturbance

The rotor angle stability is divided into small-signal and transient stability. The small-signal stability is present for small disturbances in the form of undamped electromechanical oscillations. The transient stability is due to lack of synchronising torque and is initiated by large disturbances. The time frame of angle stability is that of the electromechanical dynamics of the power system. This time frame is called the short-term time-scale, because the dynamics typically last for a few seconds.

The voltage problem is load-driven as described above. The voltage stability may be divided into short and long-term voltage stabilities according to the time-scale of load component dynamics. Short-term voltage stability is characterised by components such as induction motors, excitation of synchronous generators, and electronically controlled devices such as HVDC and Static VAR Compensator (SVC). The time-scale of short-term voltage stability is the same as the time-scale of rotor angle stability. The modelling and the analysis of these problems are similar. The distinction between rotor angle and short-term voltage instability is sometimes difficult, because most practical voltage collapses include some element of both voltage and angle instability.

When short-term dynamics have died out sometime after the disturbance, the system enters a slower time frame. The dynamics of the long-term time-scale lasts for several minutes. Two types of stability problems emerge in the long-term time-scale: frequency and voltage problem. Frequency problems may appear after a major disturbance resulting in power system islanding. Frequency instability is related to the active power imbalance between generators and loads. An island may be either under or over-generated when the system frequency either declines or rises.

The analysis of long-term voltage stability requires detailed modelling of long-term dynamics. The long-term voltage stability is characterised by scenarios such as load recovery by the action of on-load tap changer or through load self-restoration, delayed corrective control actions such as shunt compensation switching or load shedding. The long-term dynamics such as response of power plant controls, boiler dynamics and automatic generation control also affect long-term voltage stability. The modelling of long-term voltage stability requires consideration of transformer on-load tap changers, characteristics of static loads, manual control action of operators, and automatic generation control.

For the purposes of analysis, it is sometimes useful to classify voltage stability into small and large disturbances. Small disturbance voltage stability considers the power system's ability to control voltage after small disturbances, e.g. changes in load. The analysis of small disturbance voltage stability is done in steady state. In that case, the power system can be linearised around an operating point and the analysis is typically based on eigenvalue and eigenvector techniques. Large disturbance voltage stability analyses the response of the power system to large disturbances, e.g. faults, switching or loss of load, or loss of generation. Large disturbance voltage stability can be studied by using non-linear time domain simulations in the short-term time frame and load-flow analysis in the long-term time frame. The voltage stability is, however, a single problem on which a combination of both linear and non-linear tools can be used.

3.8 MECHANISM OF VOLTAGE COLLAPSE

Voltage instability, i.e. the absence of voltage stability, results in voltage collapse. It usually involves large disturbances (including rapid increase in load or power transfer) and mostly associated with reactive power deficit. Oscillatory voltage instability may also be possible. Over-voltage instability is excluded because we assume that the self-excitation of rotating machines is not normally permitted and the over-voltage in the line is normally more of an equipment problem rather than a power system stability problem.

Voltage instability and voltage collapse dynamics span a range in the time frame from a fraction of a second to the order of minutes. Table 3.2 exhibits the time frame of the components causing voltage instability. The data shows the time frame of voltage stability classified as transient and long-term (or steady state) voltage stability.

The transient voltage stability is from 0 to 10 seconds in time-scale. Its time frame is similar to rotor angle stability and sometimes may occur simultaneously. An unfavourable factor in load component (such as induction motor dynamics and dc converters) may result in voltage collapse associated with rotor angle instability. For sudden voltage dips (say slow clearing of faults), the reactive power

Table 3.2 Time responses of factors affecting voltage stability

Factors affecting transient voltage stability in time-scale	Factors affecting long-term voltage stability in time-scale
Static VAR compensator = 1 second	OLTC operation = 2 minutes
Switched capacitors = 2 seconds	Generation readjustment = 2 minutes
Generator excitation = 1.5 seconds	Line overload = 5 minutes
Induction motor dynamics = 1 second	Distribution voltage reguation = 3 minutes
Under-voltage load shedding = 10 seconds	
HVDC operation = 1–2 seconds	

demand of induction motor increases causing further voltage instability problems. In post-disturbance period, motors face difficulty in re-acceleration and may even cause stalling of adjoining motors. Electrical *islanding* and under-frequency load shedding may also cause voltage collapse particularly when the power imbalance between the areas is more than 50%. Voltage decay affects the voltage sensitive loads and slows down frequency decay and also slows down under-frequency load shedding (even the operation of under-frequency relay will be affected due to low voltage). The use of HVDC links may also affect the transient voltage stability though it may improve the transient rotor angle stability. Sometimes it may be necessary to reduce the dc power to improve the ac side voltage magnitude and security margin.

The steady state voltage stability is usually for several minutes (may be even up to 5 minutes) in time-scale. This long-term voltage stability involves usually high loads and high power imports from neighbouring areas following a large disturbance and involving high reactive power loss and voltage dips in the receiving side. Tap-changing transformers and distribution voltage transformers sense this low voltage and act to restore the distribution voltage, thus restoring load power. This load restoration causes further sag in transmission voltage. The generators are over-excited and may even be overloaded. Over-excitation limiters may act to restrict over-excitation of generators. The transmission and generation system then becomes hard-pressed to support the reactive loss. The rate of change of reactive loss with respect to decline in voltage becomes too high and this may cause rapid voltage decay. The effect is cumulative causing high demand of reactive current by induction motors, high system regulation and higher series reactive loss associated with low shunt charging. Partial or complete voltage collapse then follows.

Long-term voltage instability may also occur by an increasingly large load demand or large rapid magnitude of power transfer. This may create large reactive power demand and voltage decay starts. Any mild or moderate system disturbance may create voltage instability during this time. Timely application of reactive power equipment and strategic load shedding may save the system from voltage instability. Fast-acting reactive compensators or synchronous condensers in field forcing mode may be the choices for effective corrective actions.

3.9 ANALYSIS OF POWER SYSTEM VOLTAGE STABILITY

3.9.1 A Simple Example

The characteristics of voltage stability are illustrated by a simple example. Figure 3.4 shows a simplified two-bus test system. The generator produces active power, which is transferred through a transmission line to the load. The reactive power capability of the generator is infinite; thus the generator voltage V_1 is constant. The transmission line is presented with a reactance (jX). The load is constant power load including the active P and reactive Q parts.

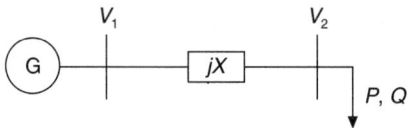

Figure 3.4 Two-bus test system.

The purpose of the study is to calculate the load voltage V_2 with different values of load. The load voltage can be calculated analytically in this simple example. Generally voltages are solved with a load-flow program. The solution of Eq. (3.10) is the load voltage for the load-flow equation of the example, when the voltage angle is eliminated.

$$V_2 = \sqrt{\frac{(V_1^2 - 2QX) \pm \sqrt{V_1^4 - 4QXV_1^2 - 4P^2X^2}}{2}} \quad (3.10)$$

The solutions of load voltage are often presented as a PV-curve (see Figure 3.5). The PV-curve presents load voltage as a function of load or sum of loads. It presents both solutions of power system. The power system has 'low current–high voltage' and 'high current–low voltage' solutions. Power systems are operated in the upper part of the PV-curve. This part of the PV-curve is statically and dynamically stable. The head of the curve is called the maximum loading point. The critical point where the solutions unite is the voltage collapse point. The maximum loading point is more interesting from the practical point of view than the true voltage collapse point, because the maximum of power system loading is achieved at this point. The maximum loading point is the voltage collapse point when constant power loads are considered, but in general they are different. The voltage dependence of loads affects the point of voltage collapse. The power system becomes voltage unstable at the voltage collapse point. Voltages decrease rapidly due to the requirement for an infinite amount of reactive power. The lower part of the PV-curve (to the left of the voltage collapse point) is statically stable, but dynamically unstable. The power system can only operate in stable equilibrium so that the system dynamics act to restore the state to equilibrium when it is perturbed.

Figure 3.5 presents five PV-curves for the test system. These curves represent different load compensation cases (tan $\phi = Q/P$). Since the inductive line losses

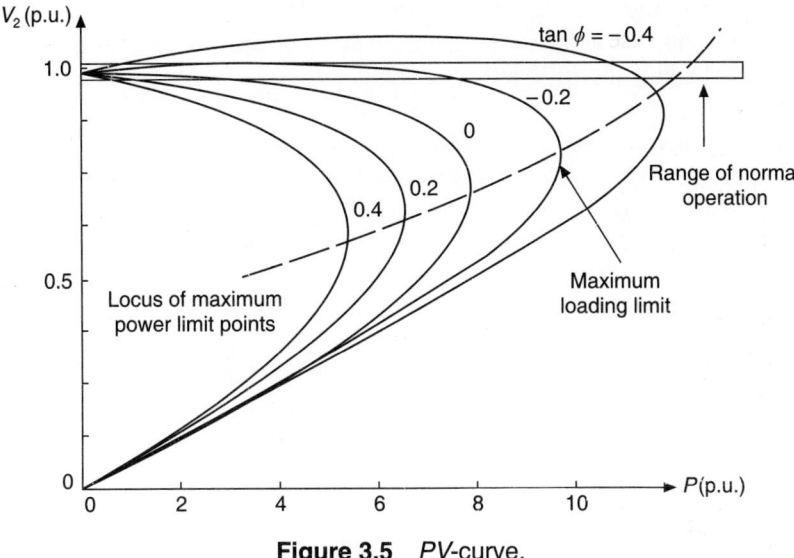

Figure 3.5 *PV*-curve.

make it inefficient to supply a large amount of reactive power over long transmission lines, the reactive power loads must be supported locally. According to Figure 3.6, addition of load compensation (decrement of the value of tan ϕ) is beneficial for the power system. The load compensation makes it possible to increase the loading of the power system according to voltage stability. Thus, the monitoring of power system security becomes more complicated because the critical voltages might be close to voltages of normal operation range.

The opportunity to increase power system loading by load and line compensation is valuable nowadays. Compensation is usually less expensive and more environment-friendly than line investments. Furthermore, the construction of new lines has become time-consuming if not even impossible in some cases. At the same time new generation plants are being constructed farther away from load centres, and more electricity is being exported (and imported). This trend inevitably requires addition of transmission capacity in the long run. The simple description of the voltage stability phenomenon has been limited to a radial system because it present a simple and a clear picture of the problem. In practical power systems, many factors affect the progress of voltage collapse due to voltage instability.

3.9.2 Analysis of Voltage Stability of Non-linear Power System

Static and dynamic analysis

Limitations of steady state power system studies (algebraic equations) are associated with stability of non-linear dynamic systems. The dynamics of the power system is modelled as differential equations. If the dynamics acts extremely fast to restore algebraic relations between the states, then it can be a good approximation to use algebraic relations. In dynamic analysis, the transition itself is of interest and it checks that the transition will lead to an acceptable operating condition.

The voltage stability is a dynamic phenomenon by nature, but the use of steady state analysis methods is permitted in many cases. Accurate dynamic simulation is needed for postmortem analysis and the co-ordination of protection and control. The voltage stability assessment of static and dynamic methods should be close to each other when appropriate device models are used and the voltage instability does not occur during the transient period of disturbance. In steady state voltage stability studies, load-flow equations are used to represent system conditions. In these studies, it is assumed that all dynamics have died out and all controllers have done their duty. Steady state voltage stability studies investigate long-term voltage stability. The results of these studies are usually optimistic compared to dynamic studies.

The advantage of using algebraic equations compared to dynamic studies is the computation speed. Dynamic simulations are time-consuming and an engineer is needed to analyse the results. However, the stability of the power system cannot be fully guaranteed with steady state studies. The time domain simulations capture the events and chronology leading to voltage instability. This approach provides the most accurate response of the actual dynamics of voltage instability when appropriate modelling is included. However, the devices which may have a key role in the voltage instability include those which may operate in a relatively long time frame. These devices include the over-excitation limiter of synchronous generator and the on-load tap changer. It may take minutes before a new steady state is reached or voltage instability occurs following a disturbance. Static analysis is ideal for the bulk of power system studies in which the examination of a wide range of power system conditions and a large number of contingencies is required.

Stability of non-linear system

The stability of a linear system can be determined by studying the eigenvalues of the state matrix. The system is stable if all real parts of eigenvalues are negative. If any real part is positive, the system is unstable. *If any real part is zero we cannot easily predict*. The stability of a non-linear system can be determined by linearisation of the system at the operating equilibrium. Using the first terms of Taylor series expansion, the function can be linearised. The state matrix is a Jacobian matrix of function determined by the state vector at the operation point. The Jacobian matrix describes the linear system which best approximates the non-linear equations close to the equilibrium. In that case, the stability of the non-linear system can be studied like the stability of linear systems in the neighbourhood of operating equilibrium.

Bifurcation analysis

Voltage stability is a non-linear phenomenon and it is natural to use non-linear analysis techniques such as bifurcation theory to study voltage collapse. Bifurcation describes qualitative changes such as loss of stability. Bifurcation theory assumes that power system parameters vary slowly and predicts how a power system becomes unstable. The change of a parameter moves the system slowly from one equilibrium to another until it reaches the collapse point. The system dynamics must act more quickly to restore the operating equilibrium than the parameter variations do to change the operating equilibrium. Although voltage collapses are typically associated with discrete events such as large disturbances, device or control limits, some useful concepts of bifurcation theory can be reused with some care. Voltage collapses often have an initial period of slow voltage decline. Later on in the voltage collapse, fast dynamics can lose their stability in a bifurcation and a fast decline of voltage results.

Bifurcation occurs at the point where, due to slow changes of a parameter, the characteristics of the system change. Bifurcation points where change from stable to unstable occurs, from stationary to oscillatory, or from order to chaos, are the most interesting point in voltage stability studies. These changes may also take place simultaneously. Usually only one parameter, e.g. load demand, is changed at once, in which case there is a possibility to achieve either saddle node or Hopf bifurcation.

At the saddle node bifurcation, the stable and unstable equilibria coalesce and disappear, then the Jacobian matrix is singular, thus one of the eigenvalues (or singular values) must be zero. The saddle node point is a limiting point between the stable and the unstable areas. The consequence of the loss of operating equilibrium is that the system state changes dynamically. The dynamics can be such that the system voltages fall dynamically. The complex conjugate eigenvalue pair is located at the imaginary axis at the Hopf bifurcation, in which case oscillation arises or disappears. The Jacobian matrix is non-singular at the Hopf bifurcation. There are also other bifurcations (e.g. pitchfork bifurcation) which are less likely to be found in general equations.

If we consider the example presented in Figures 3.5 and 3.6, the bifurcation parameter is the system loading. The system states are load voltage and angle. As the loading slowly increases, the stable and unstable solutions approach each other and finally coalesce at the critical loading.

3.9.3 Factors Affecting Voltage Stability

The purpose of the chapter is to describe the modelling of power system and to discuss the effects of power system devices and to control voltage stability. It is well known that slower acting devices such as generator over-excitation limits, the characteristics of the system loads, on-load tap changers and compensation devices will contribute to the evolution of a voltage collapse. The modelling of a power system is similar in long-term voltage stability studies than in traditional load-flow studies. Most components can be modelled with existing models. Fast-acting devices such as induction motors, excitation system of synchronous machines, control of HVDC and Static VAR Compensators (SVCs) are not described in detail here. These devices contribute to voltage stability, but the main emphasis is on short-term voltage stability. The analysis and combination of fast and slow acting devices is also difficult with traditional dynamic simulation tools, but may be easily analysed with a fast voltage stability analysis method based on quasi steady state approximation, which consists of replacing the transient differential equations with adequate equilibrium relationships.

Reactive power capability of synchronous generator

Synchronous generators are the primary devices for voltage and reactive power control in power systems. According to power system security, the most important reactive power reserves are located there. In voltage stability studies, active power and reactive power capability of generators is needed to consider accurately achieving the best results. The limits of generator active power and reactive power are commonly shown with a *PQ*-diagram (Figure 3.6). The active power limits are due to the design of the turbine and the boiler. Active power limits are constant. Reactive power limits are more complicated, which have a circular shape and are voltage dependent. Normally the reactive power limits are described as constant limits in the load-flow programs. The voltage dependence of generator reactive power limits is, however, an important aspect in voltage stability studies and should be taken into account in these studies.

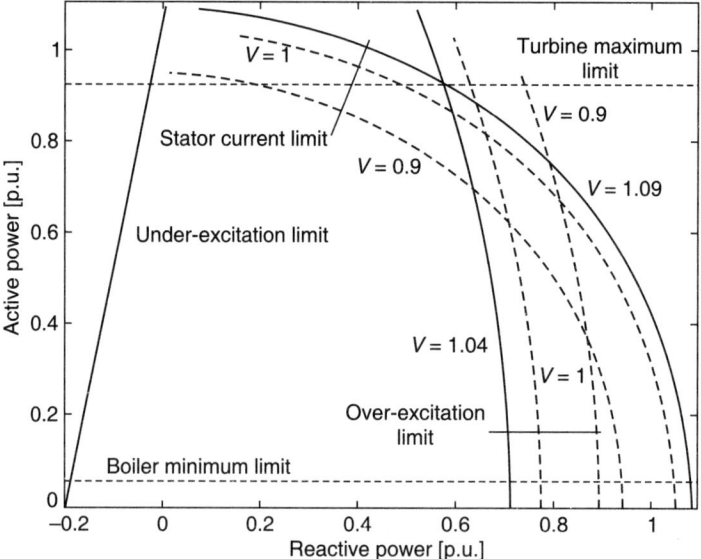

Figure 3.6 *PQ*-diagram (X_d = 0.45 p.u., I_{smax} = 1.05 p.u., E_{max} = 1.35 p.u.).

The limitation of reactive power has three different causes: stator current, over-excitation and under-excitation limits. When the excitation current is limited to maximum value, the terminal voltage is the maximum excitation voltage less the voltage drop in the synchronous reactance. The power system has become weaker, because the constant voltage has moved more remote from loads. The voltage dependent limit of excitation current can be calculated from Eq. (3.11), where P_G is the active power of generator, E_{max} is the maximum of electromotive force, X_d is the synchronous reactance and V is the terminal voltage. The reactive power limit corresponding to stator current limit can be calculated from Eq. (3.12). The terminal voltage, the maximum of stator current I_{smax} and the active power of generator P_G determine it.

$$Q_{rmax} = -\frac{V^2}{X_d} + \sqrt{\frac{V^2 E_{max}^2}{X_d^2} - P_G^2} \qquad (3.11)$$

$$Q_{smax} = \sqrt{V^2 I_{smax}^2 - P_G^2} \qquad (3.12)$$

The reactive power limits are presented in Figure 3.6 with three different values of terminal voltage: 1.04, 1.0 and 0.9 p.u. The reactive power limit in nominal voltage condition ($V = 1.04$) is defined by the excitation current. The stator current limit should be taken into account because the limit decreases when the voltage decreases. The stator current limit is more restrictive than the excitation current limit with terminal voltages of 1.0 and 0.9 p.u. when the generator output is higher than 0.8 p.u. and 0.35 p.u., respectively. Due to voltage dependence, the reactive power capability increases when the terminal voltage decreases and the generator output is less than the previous values of output power. The stator current limiter is commonly used in Sweden. It is used to limit the reactive power output to avoid stator overloading. The action of the stator current limit is disadvantageous for voltage stability. The stator current limiter decreases the reactive power capability to avoid stator overheating and causes a dramatic decrement in voltage.

Generator reactive power capability from the system point of view is generally much less than that indicated by the manufacturers' reactive capability curves, due to constraints imposed by plant auxiliaries and the power system itself. The steady operation of the power plant may be threatened when the system voltage is low. The necessary auxiliary system like pumps, fans, etc. may stop due to under-voltage, which may cause tripping of the power plant. Choosing optimum positions for the main and auxiliary transformer taps with respect to power system operating conditions may increase the reactive power capability of the generator.

Automatic voltage control of synchronous generator

It is important for voltage stability to have enough nodes where voltage may be kept constant. The automatic voltage controllers of synchronous generators are the most important for that. The action of modern voltage controllers is fast enough to keep voltage constant when generators are operated inside *PQ*-diagrams. The automatic voltage controller includes also the excitation current limiters (over- and under-excitation limiters) and in some cases also the stator current limiter.

In case of a disturbance, the excitation voltage is increased to maximum and the excitation current is increased high enough or to its maximum. The short-term maximum of excitation current is commonly twice the long-term maximum. The short-term over-loading capability is important for power system stability. Due to overheating of the excitation circuit, the excitation current must be limited after a few seconds. The heating coefficients of stator circuit are much larger than those of excitation circuit, that is why the stator current limiter is not necessary and may be done manually. The overloading capability of the generator may be improved by making the cooling of the generator more effective.

Figure 3.7 illustrates the action of the automatic voltage controller and current limiters and corresponds to the *PQ*-diagram of Figure 3.6. As the reactive power output increases, it hits the generator excitation current limit, which is dependent on the generator output power. If the reactive power output further increases, the generator loses controllability of voltage, when the operation point follows the curve of excitation current limit to the left and the output of reactive power slightly increases but the voltage decreases. The stator current limits are reached. The action of stator current limiters is dramatic. The stator current limiter decreases the reactive power capability to avoid stator overheating, which causes fast decrement in voltage. When the generator output power is increased, the action of the stator current limiter becomes more likely because the excitation and the stator current limits move closer to each other.

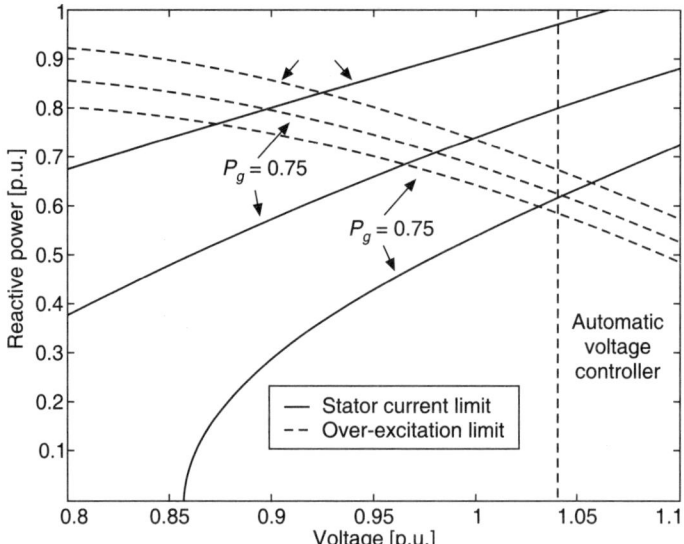

Figure 3.7 QV-diagram ($X_d = 0.45$ p.u., $I_{smax} = 1.05$ p.u., $E_{max} = 1.35$ p.u.).

Loads

The modelling of loads is essential in voltage stability analysis. The voltage dependence and dynamics of loads require consideration in these studies. Power system loads are usually represented as connected to the high voltage bus bars. The modelling of voltage dependence of loads, however, requires proper consideration of voltage control devices. An alternative and more accurate model is the representation of voltage dependent loads at the secondary side of the distribution system main transformer including a possible tap-changer control. The dynamics of loads in long-term voltage stability studies includes the operation of on-load tap changers, operation of compensation, thermostatic loads, protection system that operate due to low voltage or over-current, and extinction of discharge lamps below a certain voltage and their restart when the voltage recovers. Dynamic models for electrical motors are needed when the power system studied includes a significant amount of motor load.

The voltage dependence of loads is usually modelled with an exponent or a polynomial model. The value of the exponent describes the voltage dependence of the load. Integer values of exponents zero, one and two correspond to constant power, current and impedance loads respectively. The exponent load model is presented in Eq. (3.13), where P is active load, Q is reactive load, the subscript 0 corresponds to nominal values, V is load voltage, α is active load and B is exponent of reactive load. The typical values for the exponents for different load components are presented in Table 3.3.

$$P = P_0 \left(\frac{V}{V_0}\right)^\alpha, \quad Q = Q_0 \left(\frac{V}{V_0}\right)^\beta \tag{3.13}$$

Table 3.3 Typical values for exponents of load model

Load type	α	β
Electrical heating	2.0	$-(Q = 0)$
Television	2.0	5.2
Refrigerator/freezer	0.8–2.11	1.89–2.5
Fluorescent lighting	0.95–2.07	0.31–3.21
Frequency drives	1.47–2.12	1.34–1.98
Small industrial motors	0.1	0.6
Large industrial motors	0.05	0.5

The polynomial load model is presented in Eq. (3.14). The model sums up the voltage dependent components of integer exponent load models. The parameters of polynomial load models are Z_p, I_p and P_p for active power and Z_q, I_q and Q_q for reactive power, which describe the share of components of the total load. The measured values of parameters Z_p, I_p and P_p, and Z_q, I_q and Q_q for a refrigerator, fluorescent lighting and frequency drives are presented in Table 3.4.

$$P = P_0 \left[Z_p \left(\frac{V}{V_0}\right)^2 + I_p \left(\frac{V}{V_0}\right) + P_p \right], \quad Q = Q_0 \left[Z_q \left(\frac{V}{V_0}\right)^2 + I_q \left(\frac{V}{V_0}\right) + Q_q \right]$$

$$\tag{3.14}$$

Table 3.4 Measured values of parameters of polynomial load model

Parameter	Refrigerator/ Freezer	Fluorescent lighting			Frequency drives	
		Device 1	Device 2	Device 3	Device 1	Device 2
Z_p	1.19	0.14	0.16	0.34	0.43	3.19
I_p	−0.26	0.77	0.79	1.31	0.61	−3.84
P_p	0.07	0.09	0.05	−0.65	−0.05	1.65
Z_q	0.59	−0.06	0.18	3.03	−1.21	1.09
I_q	0.65	−0.34	−0.83	−2.89	3.47	−0.18
Q_q	−0.24	−0.60	−0.35	0.86	−1.26	0.09

The organisation of comprehensive measurements for the determination of load parameters in the whole power system is a time-consuming task. It requires measuring of load and voltage at each substation separately, covering long and various periods. The measurements should include all seasons, weather conditions, days of the week, times of day, special cases like holidays, etc. This method is used for the verification of load models and the determination of load parameters at special substations. In practice, the voltage dependence of loads is estimated with a component-based approach which builds up the load model from equipment fed from the substation. The load is first categorised into load classes: industrial, commercial, residential, agricultural, etc. The problem in this approach is the variation in values or properties of load component depending, for example, on the location of the load. This problem can be avoided if loads are categorised into classes which are typical equipment such as large motors, small motors, constant power loads, discharge lamps, thermostatic loads, compensation and distribution system voltage control. The problem in this categorisation is the difficulty to estimate the amount of load in each class. The advantage of the former method is the exact knowledge of equipment voltage dependence.

The properties of the exponent load model are presented in Figure 3.8 for the test system of Figure 3.4 ($\tan \phi = 0$). The maximum loading and the voltage collapse point of constant power load ($\alpha = 0$) is 800 MW, which corresponds to the PV-curve nose. When $\alpha = 0.7$ (Figure 3.8(a)), the maximum loading point occurs when the nominal load is about 1000 MW. The voltage collapse point is at 650 MW, when the nominal load is about 1100 MW. The nominal load corresponding to the maximum loading point is about 1300 MW, when $\alpha = 1.3$ at Figure 3.8(b).

Figure 3.9 describes the properties of the polynomial load model for the test system. The voltage dependence of Figure 3.9(a) load is about the same as the load of Figure 3.8(b). The total load consists mainly of constant impedance and current loads. Figure 3.9(b) shows an extreme case where the total load is unrealistic at the low voltage. The model is not appropriate for low voltage, because the shape of the second order polynomial is a parabola.

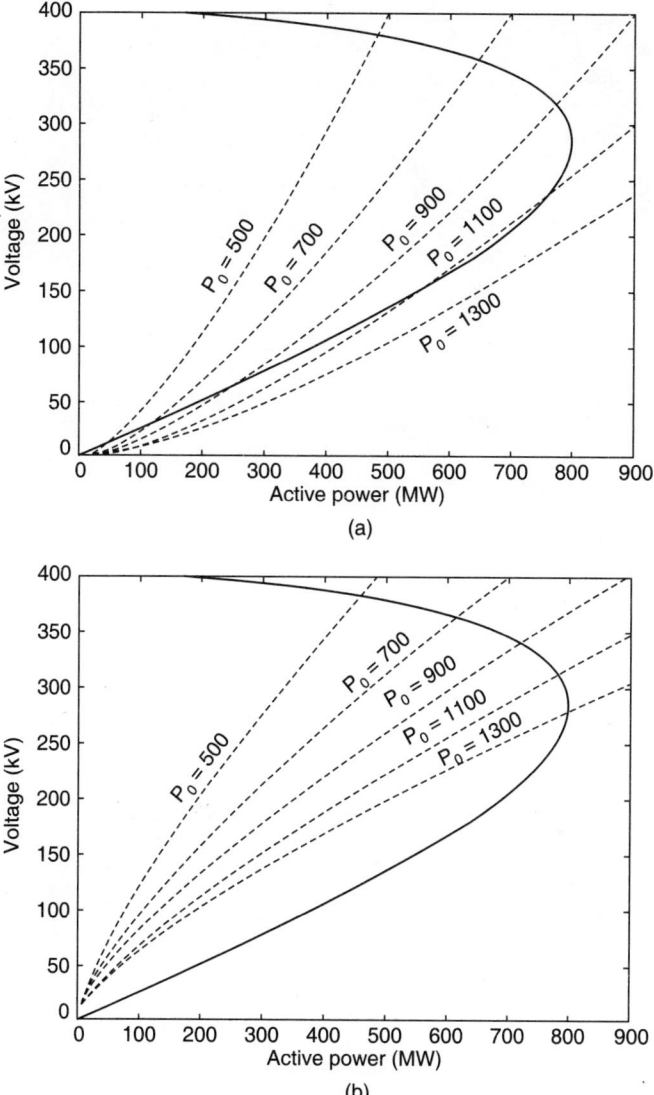

Figure 3.8 Exponent load model: (a) $\alpha = 0.7$, (b) $\alpha = 1.3$.

On-load tap changer

The automatic voltage control of power transformers is arranged with on-load tap changers. The action of the tap-changer affects the voltage dependence of load seen from the transmission network. Typically, a transformer equipped with an on-load tap changer feeds the distribution network and maintains a constant secondary voltage. When the voltage decreases in the distribution system, the load also decreases. The tap-changer operates after a time delay if the voltage error is large enough restoring the load.

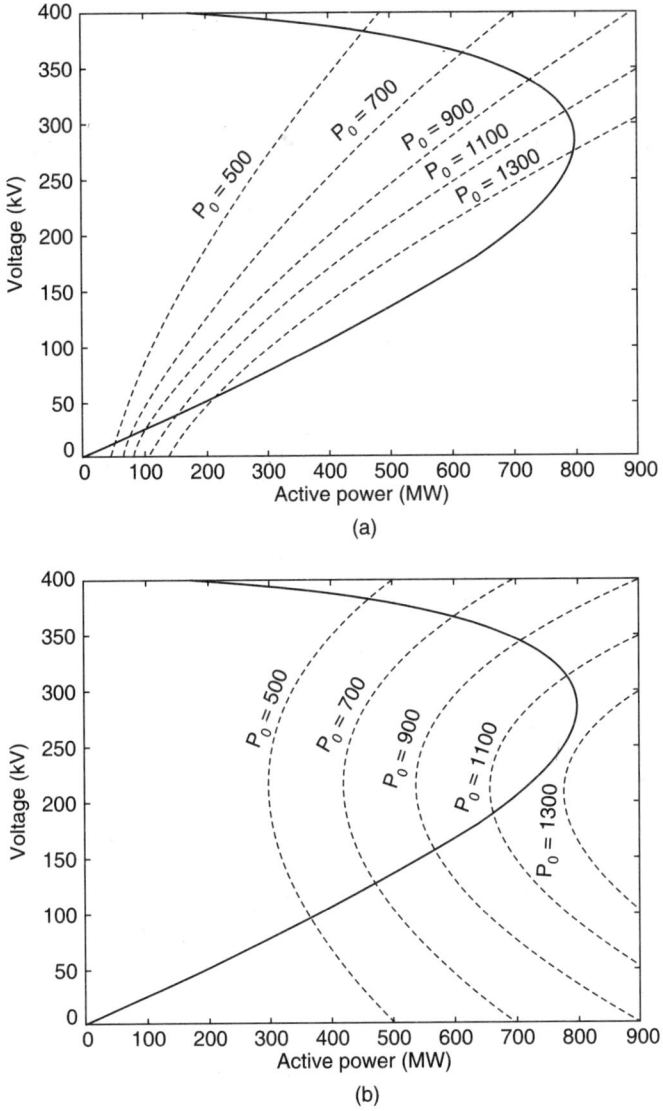

Figure 3.9 Polynomial load model: (a) $Z_p = 0.4$, $I_p = 0.5$, $P_p = 0.1$, (b) $Z_p = 1.6$, $I_p = -1.6$, $P_p = 1$.

The action of an on-load tap changer might be dangerous for a power system under disturbance. The stepping down of the tap changer increases the voltage in a distribution network; thus the reactive power transfer increases from the transmission network to the distribution network. Figure 3.10 illustrates the action of tap changer caused by a disturbance seen from the transmission network. The power system operates at point A in the pre-disturbance state. Due to the disturbance, the operation point moves to point B, which is caused by decrement of secondary voltage and load dependence of voltage. The load curve represents the state of power system just after the disturbance. After a time delay, the tap changer steps down to increase the secondary voltage. The operation point seen from the transmission network moves along the post-disturbance *PV*-curve towards a maximum loading point, which causes decrement of the primary voltage. The tap changer operates until the secondary voltage reaches the nominal voltage at point D. The amount of load at points A and D is equal due to the action of the tap-changer. The operation point D is stable, but quite close the post-disturbance maximum loading point.

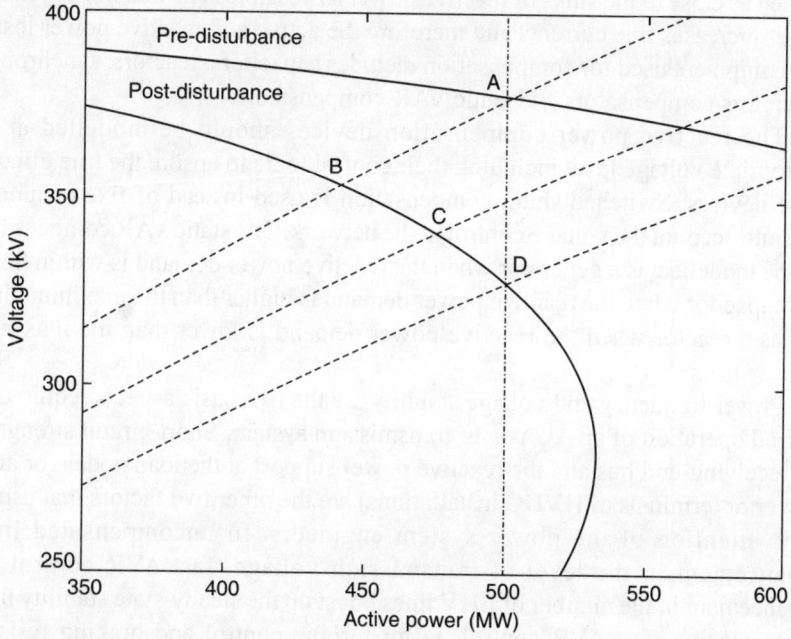

Figure 3.10 The action of no-load tap changer caused by a disturbance.

The voltage dependence of the loads can be seen when the on-load tap changer reaches the tap changer minimum limit, in which case on-load tap changer is not capable of maintaining constant secondary voltage. The step size of the on-load tap changer should also be taken into account in load-flow based long-term voltage stability studies.

The restoration of load may occur although the distribution network voltage is not increased to nominal or pre-disturbance value. A thermostat typically controls the heating and cooling loads. The energy consumed in the thermostatic loads is constant in the long run. Although heating loads are resistive, the thermostats increase the amount of load if the decrement of load voltage is long enough. The time constants of thermostatic loads are high, which makes this phenomenon slow. The thermostatic load is modelled as constant impedance load with a long time constant. A long interruption or voltage decrement might also cause a phenomenon called cold load pick-up, where the load becomes higher than the nominal value due to manual connection of additional load to compensate decreased power supply.

Compensation devices

The aim of reactive power compensation is to improve the performance of power system operation. The compensation is needed to control the reactive power balance of the power system. It is also used for minimising reactive power losses and keeping a flat voltage profile in the network. The sources of reactive power are located as close to the sinks of reactive power as possible. The transfer of reactive power increases line currents and therefore the active and reactive power losses. The equipment used for compensation includes capacitors, reactors, synchronous generators/compensators and static VAR compensators.

The reactive power compensation devices should be modelled at the appropriate voltage level including their control logic to ensure the true effect of these devices. Switched shunt compensation is used instead of fixed shunts to take into account the voltage control in the network. The static VAR compensators can be modelled as a generator when the reactive power demand is within limits, as a capacitor when the reactive power demand is higher than the maximum limit and as a reactor when the reactive power demand is lower than the maximum limit.

Power frequency and voltage stability are the two basic aspects required for desired operation of an AC power transmission system. Short-circuit strength at the receiving-end bus and the reactive power support at the load nodes (or at the convertor terminals of HVDC installations) are the other two factors that usually draw attention of the power system engineers. In uncompensated lines, enhancement in the level of transmission voltage, fast AVR control and enhancement in the number of EHV lines boost up the steady-state stability limit. Autoreclosing, fast AVR control, faster turbine control and braking resistors improve the transient stability of AC power transmission system. Though it may be difficult for uncompensated lines to improve their dynamic stability, a fast AVR control scheme may still serve as an effective tool in enhancing the dynamic stability of a power transmission line.

It has been well-established that enhancement of stability can be made by proper compensation. Shunt reactor, shunt capacitor, series capacitor, synchronous condenser, static VAR controller (SVC) and saturated reactor applications improve the steady-state system stability. The same compensation schemes, except the shunt reactor have also been found to improve the transient stability. Synchronous condenser and static VAR controller have special effect on the improvement of dynamic stability.

In practice, to counter any slow voltage change phenomenon, transformer tap setting adjustment or AVR control techniques have been adopted for uncompensated lines. Fast AVR control and fast line switching operations, including the use of breakers, improve the uncompensated system voltage conditions particularly when the system is subjected to a sudden voltage change. For compensated systems, compensating devices must be properly coordinated to produce best results when the system voltage stability is endangered. Shunt capacitor and static VAR controller improve the system voltage when the transmission system encounters any rapid or slow voltage drops. Shunt reactor, saturated reactor and static VAR controller may be used to compensate problems related to fast or slow voltage increase if there be any. These methods may be used separately or in conjunction with the tap-setting adjustment of the 'On Load Tap Changing Line Transformer' and AVR control, in case the system demands voltage control. However, line transformer tap adjustment is not generally recommended as the only tool to control the system voltage deviations in radial transmission systems, since these types of power systems are reactive power constrained and the transformer tap-setting adjustment does not alter the reactive power status. Moreover, the voltage problem is always associated with mismatch in reactive power generation and demand in these systems. For better and effective management, either the reactive power flow needs to be reoriented or load shedding is required.

In an EHV AC system, shunt reactor, synchronous condenser and static VAR controller may also serve to compensate for the high rise in voltage level due to lightning or switching. However, use of conventional protective devices must be recommended to control the sudden over-voltage phenomenon due to lightning or switching.

To enhance the short-circuit level and in order to decrease the system reactance, the service of fixed or switched series capacitors and synchronous condensers should be utilised.

We will now show that for a shunt capacitor compensated system, even if the voltage at the receiving bus and its critical value are higher than the case when the system is uncompensated, the voltage stability margin of the compensated system is less than that of an uncompensated system.

Voltage stability limit of the uncompensated line has already been discussed earlier, while that for the shunt capacitor compensated line can be obtained as follows:

Utilising the concept of static shunt capacitor compensation, the real and reactive power flow equations are given by

$$P = \frac{EV}{x} \sin \delta \tag{3.15}$$

and

$$Q = \frac{EV}{x} \cos \delta - \frac{V^2}{x} + V^2 \omega C \tag{3.16}$$

for the value of the capacitance C connected at the receiving end.

The Jacobian in this case is then given by

$$[J] = \frac{1}{x}\begin{bmatrix} (EV\cos\delta) & (E\sin\delta) \\ (-EV\sin\delta) & (-2V + E\cos\delta + 2V\omega C) \end{bmatrix} \quad (3.17)$$

With the same reasoning of singularity of the Jacobian, at voltage stability limit, the receiving-end voltage becomes

$$V_{cri} = \frac{E}{2\cos\delta\,(1 - x\omega C)} \quad (3.18)$$

Thus in a reactive power constrained line, where $Q = P\tan\phi_{net}$, utilising Eq. (3.16),

$$Q = \frac{EV}{x}\sin\delta\,\tan\phi_{net} \quad (3.19)$$

Comparing Eq. (3.16) with Eq. (3.19) and simplifying,

$$V = \frac{E\,(\cos\delta - \tan\phi_{net}\sin\delta)}{(1 - x\omega C)} \quad (3.20)$$

At voltage stability limit, comparison of Eq. (3.18) with (3.20) yields

$$\frac{E}{2\cos\delta\,(1 - x\omega C)} = \frac{E(\cos\delta - \tan\phi_{net}\sin\delta)}{(1 - x\omega C)}$$

$$\delta_{cri} = \frac{\pi}{4} - \frac{\phi_{net}}{2} \quad (3.21)$$

Thus, apparently the critical power angle remains the same at voltage stability limit for uncompensated as well as for shunt capacitor compensated lines. Observation reveals that in case of shunt capacitor compensated lines, δ_{cri} increases in magnitude at voltage stability limit, due to improvement of the load-end power factor, thus enhancing the power transfer capability. However, the receiving-end voltage magnitude at voltage stability limit differs considerably for uncompensated and shunt compensated line.

Thus,

$$V_{cri\,(comp.)} = \frac{E}{\cos\delta} \cdot K$$

where

$$K = \frac{1}{(1 - x\omega C)}, \quad K \text{ being always greater than unity.}$$

A concept of voltage stability margin (M_{vst}) is utilised at this stage and is given by

(M_{vst}) = [(rated voltage of receiving end in p.u. − voltage of receiving end in p.u. at stability limit)/rated voltage of receiving end in p.u.]

$$= \left(1 - \frac{V_{cri}}{V_{rated}}\right) \text{ in p.u.}$$

$$= \left(1 - \frac{V_{cri}}{V_{rated}}\right) \times 100\%$$

It is interesting to compare the voltage stability margins for the two cases (uncompensated and fixed capacitor compensated) represented by

$$M_{v(\text{uncomp.})} = \left[1 - \frac{V_{cri\,(\text{uncomp.})}}{V_{rated}}\right] \times 100\% \qquad (3.22)$$

$$M_{v(\text{comp.})} = \left[1 - \frac{V_{cri\,(\text{uncomp.})} \cdot K}{V_{rated}}\right] \times 100\% \qquad (3.23)$$

where $K > 1$.

A comparison of Eq. (3.22) with Eq. (3.23) reveals that

$$M_{v\,(\text{uncomp.})} > M_{v\,(\text{comp.})}$$

The receiving end voltage at voltage stability limits have been investigated for an uncompensated line at varying power factors. The same exercise is repeated for a shunt capacitor compensated model. Figure 3.4 exhibits the results graphically to show the effect of shunt capacitance on the critical receiving-end voltage profile at stability limit and on the voltage stability margin.

3.10 VOLTAGE COLLAPSE AND MODELLING OF VOLTAGE COLLAPSE

The main symptoms of voltage collapse are: low voltage profile, shortage of reactive power and heavily loaded systems. The consequences of collapse often require long system restoration, while consumers are left without supply for a long period of time. Schemes which mitigate against collapse need to use the symptoms to diagnose the approach of the collapse in time to initiate corrective action. In order to determine the symptoms, modelling of voltage collapse is important.

3.10.1 Modelling of Voltage Collapse

Modelling techniques can be divided into two parts: static or dynamic. For determining the suitability of different approaches, it is important to study the various events which affect the speed and probability of voltage collapse.

Event 1: *Disturbance of topology:* Equipment outages or faults followed by equipment outage are the examples of disturbance of topology. Generally these type of disturbances are associated with transient stability. For analysis of these events, a dynamic system model is required.

Event 2: *Load disturbance:* Load disturbance can be split into two parts: slow load fluctuation and fast load fluctuation. For slow load fluctuation, a static system model is more useful than a dynamic model.

There are some other types of disturbances which require dynamic analysis, and are the causes of voltage instability. They are:

(i) The post-disturbance equilibrium that has a low voltage profile.
(ii) The transient voltage dips that occur during the disturbance and are too long.
(iii) The post-disturbance equilibrium that is voltage unstable.

But the time domain simulation is the best way to identify all the aspects of transient and/or steady performance of the system before, during and after the disturbance.

3.11 REAL AND REACTIVE POWER OPERATING CONTOUR OF A RADIAL POWER TRANSMISSION SYSTEM

In this section, analytical tools have been developed to obtain the real and reactive power operating contour of a basic transmission system on the two dimensional complex S-plane, under a set of specified operating criteria and at different receiving-end voltage magnitudes. Later, the results obtained from such an analysis of a 400 kV LPS (longitudinal power system) line have been graphically plotted.

In the equivalent π line model, the shunt capacitances of the line have been assumed to be lumped and connected at the two extreme ends of the line, i.e. the sending and receiving ends. The entire series reactance of the system has been assumed to be connected between the two buses utilising the lumped parameter concept. The system may now be considered to be a two area power network with the EHV transmission line transporting power from the source side to the load side. The terminal reactances have been assumed to be included in the series reactance between the buses, and the mutual induction effect between the conductors have been neglected.

The receiving-end current in this equivalent line model is given by

$$I_R = (S/V)^* \qquad (3.24)$$

Also, the system admittance $[Y]$ is given by

$$[Y] = \begin{bmatrix} Y_{RR} & Y_{RS} \\ Y_{SR} & Y_{SS} \end{bmatrix} = \begin{bmatrix} (Y_L + Y_{SH}) & (-Y_L) \\ (-Y_L) & (Y_L + Y_{SH}) \end{bmatrix}$$

Utilising the conventional expression $[I] = [Y][V]$, this system can be represented by

$$\begin{bmatrix} I_R \\ I_S \end{bmatrix} = \begin{bmatrix} Y_{RR} & Y_{RS} \\ Y_{SR} & Y_{SS} \end{bmatrix} \begin{bmatrix} V \\ E \end{bmatrix} \qquad (3.25)$$

i.e.
$$I_R = Y_{RR} \cdot V + Y_{RS} \cdot E \qquad (3.26)$$

Comparing Eq. (3.24) with Eq. (3.26),

$$\left(\frac{S}{V}\right)^* = Y_{RR} \cdot V + Y_{RS} \cdot E$$

or
$$S^* = Y_{RR} \cdot V^2 + Y_{RS} \cdot V^* \cdot E$$

or
$$\left(\frac{S^*}{Y_{RR}}\right) = V^2 + \left(\frac{Y_{RS}}{Y_{RR}} \cdot E\right) \cdot V^*$$

$$\left(\frac{S^*}{Y_{RR}}\right) = V^2 + v V^* \tag{3.27}$$

where $v = \left(\dfrac{Y_{RS}}{Y_{RR}} \cdot E\right)$.

On rearrangement, Eq. (3.27) becomes

$$S - Y_{RR}^* \cdot V^2 = v^* \cdot Y_{RR}^* \cdot V \tag{3.28}$$

Considering the magnitudes only, Eq. (3.28) becomes

$$|S| - |Y_{RR}^* \cdot V^2| = |v^* \cdot Y_{RR}^* \cdot V| \tag{3.29}$$

Equation (3.29) represents a circle with $(Y_{RR}^* \cdot V^2)$ as centre and $|v^* \cdot Y_{RR}^* \cdot V|$ as its radius. Hence all states having the constant amplitudes of (v) lie on circles with these parameters on the S-plane. Each circle represent the locus of S, the receiving end complex power, for any stable value of V, the receiving-end voltage.

A typical 400 kV transmission system has been assumed for simulation which consists of a single circuit power line with bundle ACSR conductors on H type structure. The sending-end voltage E has been assumed to be constant at 1.00 p.u. The results have been plotted graphically in Figure 3.11, where circles represent the operating power contour of this particular power system. The variation of load-end voltage has been seen to affect the load-end power profile and indicates the necessity of proper voltage control at the load bus.

The investigation revealed that the complex operating power in a reactive power-constrained EHV system is a circular contour. Each circular locus represents the maximum magnitudes of real and reactive power at a specified sending-end voltage and at any fixed value of the receiving-end voltage. Any improvement in the receiving end voltage enhances the power transfer capability of the line, particularly the reactive power transfer capability.

Though a large number of circular contours can be plotted utilising the above concept, only the operating contours of the receiving-end voltage varying within the range of 0.95 p.u. to 1.05 p.u. are feasible in any practical power system.

3.12 BASIC ASPECTS OF VOLTAGE STABILITY

The concept of dynamic voltage stability is directly related to the concept of steady-state voltage stability. The new concept of dynamic voltage stability is basically developed to assess the stability of voltage being influenced by dynamic factors. To understand this concept, let the basic scenario of network operation

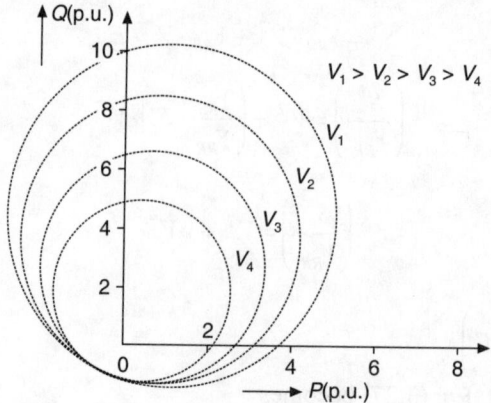

Figure 3.11 Real and reactive power contour of a transmission line with different receiving-end voltages.

that induces a state of voltage instability be considered. Let '0' denote the given initial state of operation when the system is 'stable'. It has been explained earlier that dynamic stability indicates the overall stability of a system and is achieved when all the constituent stabilities (e.g. real power steady state, transient state, etc.) are maintained. The following discussion has been made assuming the problem with the voltage stability only.

With the simple illustration of the radial type line, Figure 3.12 represents the receiving-end voltage (V) vs real power flow profile. The corresponding power and voltage limits have also been indicated in the figure. In the scenario under consideration, there is a maximum amount of power (both real and reactive) that can be transmitted to the receiving-end bus. Corresponding to this power level, the critical receiving-end voltage is given by V_{cri} [the expressions for V_{cri}, P_{max}, and δ_{lim} have been derived earlier]. It can be observed that when $V > V_{cri}$ the operation is satisfactory—statically—provided the power supplied to the load bus increases as soon as the conductance of the load is increased dynamically, provided the conventional temporal law of power variation in the network holds good (i.e. for small perturbations), and there is always an equilibrium point finally attained by the system.

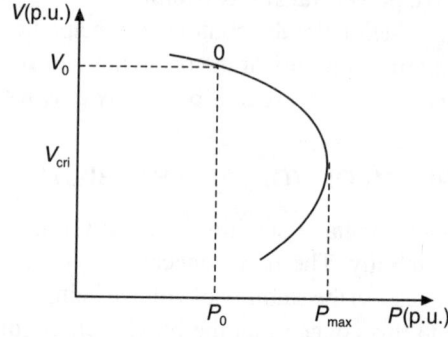

Figure 3.12 V-P profile for a radical power line.

On the other hand, if $V < V_{cri}$, an increase in the load conductance leads to a fall in the power output, and the regulation of on-load tap-changers is not stable. The state(s) corresponding to $V < V_{cri}$ is not stable. Obviously during stable operations, the differences ($\Delta P = P_{max} - P_0$ and $\Delta Q = Q_{max} - Q_0$) represent the active and reactive power margins that are liable to be supplied at each load bus from state '0' just before the concerned bus becomes 'voltage unstable'.

Concerning the dynamic aspect of operation, it may be observed that the trajectory followed between '0' and limiting state depends on the operation of a complete set of regulators and governing systems. These regulators and governing systems are dependent at the local or regional levels on the value of certain variables and on specific laws of control. The actions taken by these control equipment are interdependent. Hence for proper analysis, the system dynamics should be taken into account.

The difference of the operating state between the final and the initial position depends on various factors such as:

(a) The control law governing the compensation of reactive power at the load bus
(b) Load response
(c) Dynamic characteristic of the tap changers
(d) Existence of secondary voltage control
(e) Action taken by the operator.

The dynamic aspect is mostly affected by the factor where the system's reactive power limitations exist, and where the operation of protective systems when the generator is delivering the reactive power to a reactive power-constrained bus and the reactive power loading capability of the generator mostly dictate the trajectory of the curve shown in Figure 3.12.

With all these contributing factors, the profile of change in voltage as a function of active and reactive loads is highly nonlinear and difficult to forecast. Figures 3.13(a) and 3.13(b) represent two cases when these dynamic factors influence the trajectory of the voltage profile.

In both the cases, it may be observed that dynamic factors are involved in an adverse way resulting in the reduction of the margin of power flow (in these figures, the reduction in real power margin has been shown only, which, for reactive power, will be identical). Hence, it is evident that the dynamic factors may adversely affect and cripple the system at a load level which under a contingency is much lower than the estimated value. Actual assessment depends upon the given state of the network, the trend of voltage variation, and available reserves. Load behaviour and the use of reactive power compensation equipment should also be taken into account.

3.13 VOLTAGE SECURITY

Most voltage collapse incidents have followed from a major disturbance such as loss of a transmission line or generator and loss of reactive power compensating

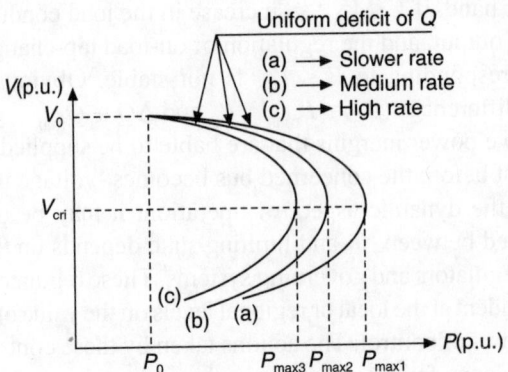

Figure 3.13(a) Reduction of maximum power transfer for system voltage profiles (Assessed Profiles).

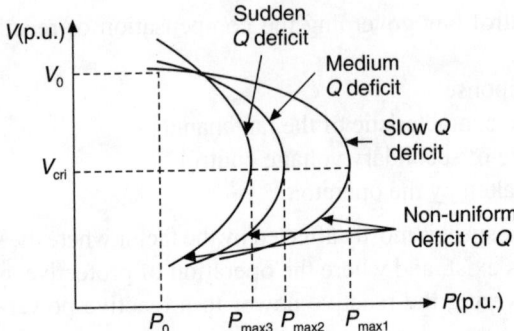

Figure 3.13(b) Voltage profiles with dynamic reactive power shortages (Actual Profiles).

device. However, as has been mentioned earlier, voltage instability, in most cases took place slowly. Investigations reveal that at the very beginning, when the network encounters a contingency, it suffers from a voltage dip which may mostly be recovered by increasing the reactive power output of the generator or by reducing the load level if the system is reactive power constrained. However, with enhancement in the voltage level, the natural trend is to enhance the loading and to relieve the generator from being overloaded and to reduce the excitation. It has been discussed earlier how would the above factors contribute towards the decrease in network voltage. This may lead to a sustained monotonic voltage decrease, leading finally to the loss of system integrity.

As dynamic factors affect the profile of voltage vs power curve, until proper analysis is made, the results may be a poor prediction of the margins that are very important in deciding the limit of stability. This leads to the concept of voltage security which actually tells the operator about the margin in real and/or reactive powers, required to have a stable voltage state following any perturbation. The voltage security is thus governed by the respective margins ΔP and ΔQ where the system is influenced by the dynamic operating factors. Here, in Figure 3.14, it has been indicated that the voltage security is actually the margin between Q_{max2} and Q_0 and not between Q_{max1} and Q_0.

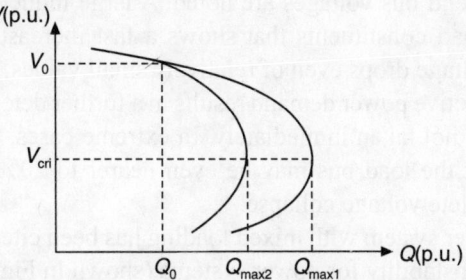

Figure 3.14 Voltage profiles to explain voltage security.

Hence, the voltage security level of the network is required to be established to assure the consumers a reasonable availability of the energy supply. An indication of the security level, particularly for reactive power reserve conditions, can be given as the distance to voltage instability (i.e. the critical voltage). To investigate this distance, multiple load flow solutions could be carried out. However, as the network becomes nearer to the condition of voltage instability, divergence in load flow calculations may occur. This is due to either numerical ill-conditioning or due to the fact that voltage instability condition has already been reached. This may lead to operational problems in the application of the numerical techniques to obtain voltage security. A better method is to use sensitivity $[dV/dQ]$ to assess the voltage collapse point. This can be applied to each and every bus of the network concerned, and the voltage security can then be calculated for these buses. The voltage security is assessed by detecting the sensitivity above or below the critical voltage, i.e. by observing the signs and slopes of the corresponding sensitivities.

3.14 TRANSIENT VOLTAGE STABILITY

The previous discussions reveal that the phenomenon of voltage instability is characterised by a progressive fall in voltage magnitude at a particular location or at a particular area, and may finally spread out in the entire network causing system voltage collapse. This phenomenon may be attributed to the inability of the power system to meet a certain load demand of reactive power.

It has already been shown in the previous chapters that voltage instability may take several forms, primarily depending upon the nature and characteristics of the loads connected and the operational state of the power transmission system. It is also governed by the dynamics of the reactive power injection components in the network.

Voltage instability being primarily a steady state phenomenon as described in the earlier discussions, transient voltage instability has also been observed in recent years. It is characterised by a sharp and sudden fall in system voltage and is possibly governed by the situation of the load bus when the system experiences voltage swings, and by the dynamics of induction motor loads. In the state of transient voltage instability, uncontrolled voltage oscillations and rapid decline in load or receiving-end bus voltages are noted. A large induction motor is one of the commonest load constituents that shows a fast increase in reactive power demand due to voltage drops even of relatively small values. This enormous and fast increase in reactive power demand results in a further deterioration of voltage, if proper action is not taken immediately. In extreme cases, the deterioration of the bus voltage at the load bus may be even nearer to 20% of the rated value indicating a complete voltage collapse.

A simple power system with mixed loading has been cited for simulations of transient voltage instability for power systems (shown in Figure 3.15). The total load composition at load bus has been assumed to be 50% induction motor loads and 50% static impedance loads.

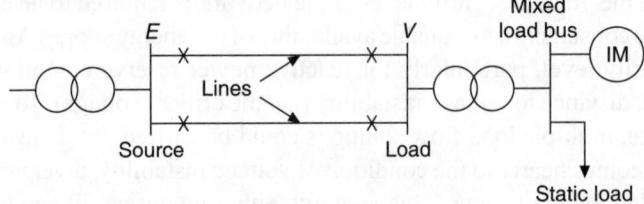

Figure 3.15 A simple power system with mixed loads.

The system has been disturbed by tripping out one circuit of the transmission system. Figure 3.16 shows the system response for one second after the disturbance. It may be observed that the decline in voltage is sharp till voltage collapse takes place. The simulation also shows the dynamic characteristics of real and reactive power demand at the bus, assuming the induction motor only as well as assuming the composite load.

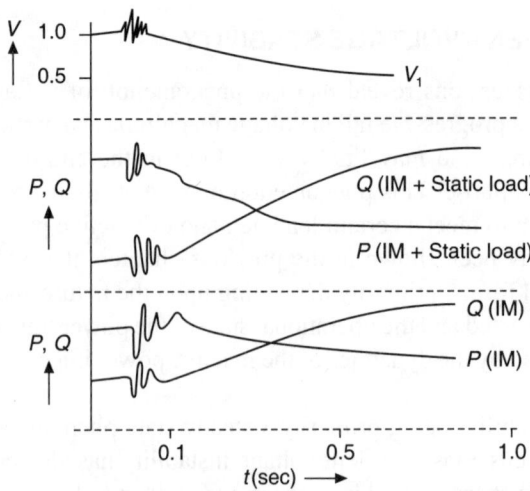

Figure 3.16 Dynamic condition of V, P, Q with contingency.

However, for the same initial total loading and system disturbance, with the assumption of a pure static impedance load with constant impedance, the system response has been shown in Figure 3.17. Here, it has been shown that this condition, for the static impedance loads, does not invite transient voltage instability.

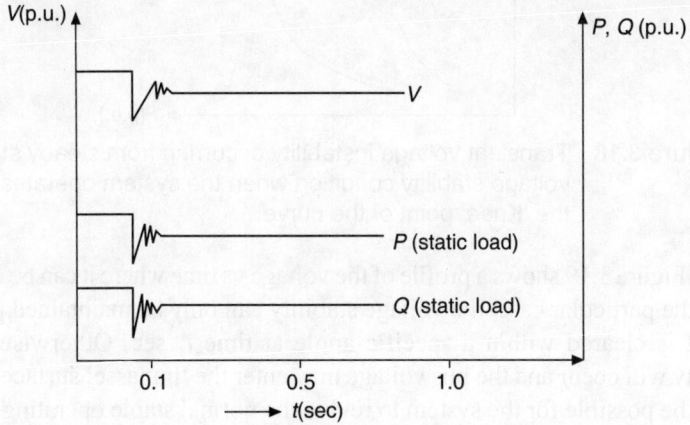

Figure 3.17 Dynamic condition of P, Q and V with contingency for static impedance loads only.

Transient voltage instability may take place in the system when the system is operating on the voltage profile near its steady state voltage stability limit. Figure 3.18 indicates the conventional V–P characteristic where the most probable condition of voltage instability will be nearer to the knee of the curve. In case of sudden deficit in reactive power, the system may face transient voltage instability. This has been shown by the dotted line in Figure 3.18 where the sudden reactive power crisis has been assumed to be due to the tripping of the EHV lines in the original system (Figure 3.15). It may be noted, in this context, that the transient voltage instability will be primarily due to the governing effect of the reactive power shortage as well as due to the dynamic characteristic of the induction motor at the load bus, which sharply influences the reactive power demand in an adverse way. Had there not been any induction motor, the transient voltage instability could have been prevented.

It has been also suggested in the literature that the factor ($\Delta Q/\Delta P$) can be used as a well-defined measure of system degradation. ΔQ is the amount of reactive power (net) to be injected to the system to maintain the terminal voltage constant due to increase in load given by ΔP. It has been suggested that the ratio should not exceed 0.9, i.e. for an additional increase of 1 MW transmitted power to maintain the terminal voltage constant, 0.9 MVAR must be injected. This value however reflects the limit beyond which shunt compensation or reactive power injection is no longer suitable and a new transmission line is required.

Researchers stressed on the role of the critical clearing angle on the event (say fault) that invites voltage instability problems. It has been shown that the critical clearing angle of the generator has an important implication on the voltage

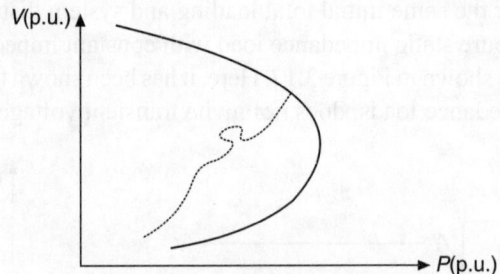

Figure 3.18 Transient voltage instability occurring from steady state voltage stability condition when the system operates near the 'Knee' point of the curve.

stability. Figure 3.19 shows a profile of the voltage vs time where it can be observed that for the particular case, the voltage stability can only be maintained provided the fault is cleared within a specific angle at time t_1 sec. Otherwise voltage instability will occur and the bus voltage may enter the 'impasse' surface where it may not be possible for the system to revive the normal stable operating voltage, even if the fault is cleared.

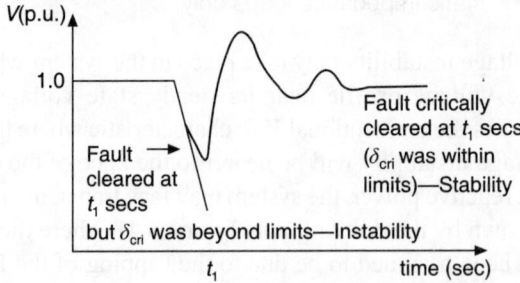

Figure 3.19 Role of critical clearing angle on voltage stability.

The role of induction motors has also been highlighted in that reference which leads to the conclusion that, for the induction motor predominant load buses, the critical clearing time is important to ensure stable operating voltage. Till the system state does not enter the 'impasse' surface, it is possible to regain the bus voltage following the fault clearing. Once the system voltage enters into that zone, any attempt to restore the normalcy of bus voltage does not serve the purpose of maintaining voltage stability.

Hence from the above discussions, it is evident that the transient voltage instability depends not only on the load behaviour and system reactive power limitation, but also on the critical clearing angle of the generator and critical clearing time for the system to regain voltage stable state.

EXERCISES

1. What are the salient disturbances that cause voltage instability?
2. Briefly discuss the concept of voltage stability from the viewpoint of reactive power deficit.
3. What is the relation of voltage stability with rotor angle stability?
4. What do you mean by stability margin? How would you develop the criteria of steady state static and transient reactive power stability? Also define the voltage stability margin.
5. Classify power system operation from the viewpoint of voltage instability, voltage collapse and short and long term stability.
6. Discuss the mechanism of voltage collapse in a power transmission system.
7. How do you get the magnitude of load voltage mathematically from load flow equation? What are the implications?
8. How does the bifurcation analysis help in determining the voltage stability limit?
9. Briefly discuss the factors affecting voltage stability.
10. Describe qualitatively the role of tap changer in governing the condition of voltage instability.
11. How does a reactive power compensation device help in maintaining the voltage stability of a transmission system? Compare stability margins of uncompensated systems with those of compensated systems.
12. Develop the real and reactive power operating contour of a radial power transmission system.
13. What is voltage security? Explain its implications on power system operation.
14. What do you mean by transient voltage instability? What is the role of critical clearing angle on voltage instability?

Chapter **4**

Power Transfer at Voltage Stability Limit

4.1 INTRODUCTION

The fundamental concepts of voltage stability having been introduced in an earlier chapter, it is now desired to present the aspects of power transfer at the voltage stability limit of a simple power line model. The aim is to develop the magnitude of the receiving-end bus voltage at the limit of voltage stability and to highlight the role of reactive power in maintaining the voltage stability limit. The load flow solution is also closely associated with voltage stable operation of the line and the loading limit of a line is also governed by the voltage stability limit.

4.2 MAGNITUDE OF RECEIVING-END VOLTAGE

At voltage stability limit

A basic line model has been assumed when a large area of generation 'A' feeds a concentrated load centre 'B' having little generation through a radial line. The receiving-end bus voltage is assumed to be uncontrolled and the power transfer takes place from A to B though this reactive power-constrained line.

The power flow equation at the receiving-end bus is given by

$$S_R = P_R + jQ_R \qquad (4.1)$$

where P_R and Q_R are given by

$$P_R = \frac{EV}{B}\cos(\beta - \delta) - \frac{A}{B}V^2\cos(\beta - \alpha) \qquad (4.2)$$

$$Q_R = \frac{EV}{B} \sin(\beta - \delta) - \frac{A}{B} V^2 \sin(\beta - \alpha) \tag{4.3}$$

Assuming the sending-end voltage to be constant, the receiving end powers become the functions of V and δ. Hence, Eqs. (4.2) and (4.3) can also be represented in terms of V and δ is given by

$$P_R = f_1(\delta \text{ and } V) \tag{4.4}$$

$$Q_R = f_2(\delta \text{ and } V) \tag{4.5}$$

The Jacobian of these power flow equations is then given by

$$J = \begin{bmatrix} \dfrac{\partial f_1}{\partial \delta} & \dfrac{\partial f_1}{\partial V} \\ \dfrac{\partial f_2}{\partial \delta} & \dfrac{\partial f_2}{\partial V} \end{bmatrix}$$

$$= \frac{1}{B} \begin{bmatrix} -EV \sin(\delta - \beta) & E \cos(\delta - \beta) - 2AV \cos(\beta - \alpha) \\ -EV \cos(\delta - \beta) & -E \sin(\delta - \beta) - 2AV \sin(\beta - \alpha) \end{bmatrix} \tag{4.6}$$

The determinant of the matrix in Eq. (4.6) is given by

$$\Delta[J] = \frac{1}{B}[E^2 V - 2EV^2 A \cos(\delta - \alpha)] \tag{4.7}$$

The Jacobian is singular when $\Delta[J] = 0$, signifying the single solution of basic load flow equations. The point of voltage stability limit being coincident with the singularity of the Jacobian [1, 2], the condition of voltage stability limit may be obtained from the equation $\Delta[J] = 0$ which yields

$$V = \frac{E}{2 \cos(\delta - \alpha) A}$$

Neglecting α and assuming $A = 1$,

$$V = \frac{E}{2 \cos \delta} \tag{4.8}$$

Equation (4.8) represents the *magnitude of the receiving-end voltage at voltage stability limit*.

During maximum power transfer

From Eqs. (4.2) and (4.3),

$$P_R + jQ_R = \left[\frac{EV}{B} \cos(\delta - \beta) - \frac{A}{B} V^2 \cos(\beta - \alpha) \right]$$

$$+ j \left[-\frac{EV}{B} \sin(\delta - \beta) - \frac{A}{B} V^2 \sin(\beta - \alpha) \right]$$

or
$$\left[P_R + \frac{A}{B}V^2 \cos(\beta - \alpha)\right] + j\left[Q_R + \frac{A}{B}V^2 \sin(\beta - \alpha)\right]$$
$$= \frac{EV}{B} \cos(\delta - \beta) - j\frac{EV}{B} \sin(\delta - \beta)$$

or
$$\left[P_R + \frac{A}{B}V^2 \cos(\beta - \alpha)\right]^2 + \left[Q_R + \frac{A}{B}V^2 \sin(\beta - \alpha)\right]^2$$
$$= \frac{E^2 V^2}{B^2} \quad (4.9)$$

Substituting
$$\frac{A}{B} \cdot \cos(\beta - \alpha) = a$$
$$\frac{A}{B} \cdot \sin(\beta - \alpha) = -b$$

and
$$\frac{1}{B} = C$$

Eq. (4.7) can be modified as
$$(P_R + aV^2)^2 + (Q_R - bV^2)^2 = C^2 E^2 V^2 \quad (4.10)$$

However, in a radial power line, the magnitude of reactive power flow is restricted and the reactive power available at the receiving bus is determined solely by the power factor of the load end and is given by
$$Q_R = P_R \tan \phi \quad (4.11)$$

Hence, from Eq. (4.10), substituting the value of Q_R from Eq. (4.11), we obtain
$$(P_R + aV^2)^2 + (P_R \tan \phi - bV^2)^2 = C^2 E^2 V^2$$

or $P_R^2 + a^2 V^4 + 2 P_R aV^2 + P_R^2 \tan^2 \phi + b^2 V^4 - 2 P_R \tan \phi \cdot bV^2 = C^2 E^2 V^2$

or $\{P_R^2 (1 + \tan^2 \phi) + 2 P_R \sec \phi \cdot aV^2 \cos \phi - 2 P_R \sec \phi \cdot \sin \phi \cdot bV^2\}$
$$+ V^4 (a^2 + b^2) = C^2 E^2 V^2$$

or $P_R^2 \sec^2 \phi + 2 P_R \sec \phi \cdot aV^2 \cdot \cos \phi - 2 P_R \sec \phi \cdot bV^2 \sin \phi$
$$+ V^4 (a^2 + b^2) = C^2 E^2 V^2$$

or $[P_R \sec \phi + V^2 (a \cos \phi - b \sin \phi)]^2 + V^4 (a \sin \phi + b \cos \phi)^2 = C^2 E^2 V^2$
$$(4.12)$$

The condition for maximum transmitted power in this reactive power-constrained line may be found by differentiating Eq. (4.12) with respect to V and then putting

$(dP_R/dV) = 0$ in the derived equation. This subsequently gives

$$P_R \sec\phi + (a\cos\phi - b\sin\phi)V^2 = \frac{C^2E^2 - 2(a\sin\phi + b\cos\phi)^2 V^2}{2(a\cos\phi - b\sin\phi)}$$

(4.13)

From Eqs. (4.12) and (4.13),

$$\left\{\frac{C^2E^2 - 2(a\sin\phi + b\cos\phi)^2 V^2}{2(a\cos\phi - b\sin\phi)}\right\}^2$$

$$= C^2E^2V^2 - V^4(a\sin\phi + b\cos\phi)^2$$

or $\quad \dfrac{\{C^4E^4 + 4(a\sin\phi + b\cos\phi)^4 V^4 - 4C^2E^2(a\sin\phi + b\cos\phi)^2 V^2\}}{4(a\cos\phi - b\sin\phi)^2}$

$$= C^2E^2V^2 - V^4(a\sin\phi + b\cos\phi)^2$$

or $\quad C^2E^2 \cdot 4(a\cos\phi - b\sin\phi)^2 V^2 + 4C^2E^2V^2(a\sin\phi + b\cos\phi)^2$

$$- 4(a\cos\phi - b\sin\phi)^2 V^4 (a\sin\phi + b\cos\phi)^2$$

$$- 4(a\cos\phi + b\sin\phi)^4 V^4 - C^4E^4 = 0$$

or $\quad 4C^2E^2V^2(a^2+b^2) - 4V^4(a\sin\phi + b\cos\phi)^2$

$$[(a\cos\phi - b\sin\phi)^2 + (a\sin\phi + b\cos\phi)^2] - C^4E^4 = 0$$

or $\quad -4V^4(a\sin\phi + b\cos\phi)^2(a^2+b^2) + 4C^2V^2E^2(a^2+b^2) - C^4E^4 = 0$

which is a quadratic equation in V^2.

Solving for V^2, we obtain

$$V^2 = \left[-4C^2E^2(a^2+b^2) \pm \right.$$

$$\left. \frac{\sqrt{16C^4E^4(a^2+b^2)^2 - 16(a^2+b^2)(a\sin\phi + b\cos\phi)^2 C^4E^4}}{-8(a^2+b^2)(a\sin\phi + b\cos\phi)^2} \right]$$

$$= -\frac{C^2E^2}{-2(a\sin\phi + b\cos\phi)^2} \pm$$

$$\frac{C^2E^2\sqrt{(a^2+b^2) - (a\sin\phi + b\cos\phi)^2}}{-2\sqrt{(a^2+b^2)}(a\sin\phi + b\cos\phi)^2}$$

$$= \frac{\left[\sqrt{(a^2+b^2)} \mp \sqrt{(a\sin\phi + b\cos\phi)^2 + (a\cos\phi - b\sin\phi)^2 - (a\sin\phi + b\cos\phi)^2}\right]}{2\sqrt{(a^2+b^2)}(a\sin\phi + b\cos\phi)^2} C^2 E^2$$

(4.14)

$$\therefore \quad V = \pm \sqrt{\left[\frac{(a^2+b^2)^{1/2} - (a\cos\phi - b\sin\phi)}{2(a^2+b^2)^{1/2}(a\sin\phi + b\cos\phi)^2}\right]} CE$$

Equation (4.14) represents the receiving-end voltage for the condition of maximum power transfer when the reactive power constraint in the line exists, and when the receiving-end reactive power is governed by the load power factor. Considering the positive sign of V in Eq. (4.14) only (as the negative voltage magnitude does not carry any physical meaning) and also as

$$\sqrt{a^2 + b^2} = \frac{A}{B}$$

$$(a\cos\phi + b\sin\phi) = \frac{A}{B}\cos(\phi - \beta + \alpha)$$

$$(a\sin\phi + b\cos\phi) = \frac{A}{B}\sin(\phi + \beta - \alpha)$$

from Eq. (4.14), we can write the expression for receiving-end voltage as

$$V = \left[\frac{(A/B) - (A/B)\cos(\phi - \beta + \alpha)}{2(A/B)\{(A/B)\sin(\phi + \beta - \alpha)\}^2}\right]^{1/2} \cdot CE \qquad (4.15)$$

In the lossless frame, $\alpha = 0$ and $\beta = 90°$
∴ Eq. (4.15) becomes

$$V_{max} = E\left[\frac{1 - \sin\phi}{2A^2\cos^2\phi}\right]^{1/2} = E\left[\frac{1}{2(1+\sin\phi)}\right]^{1/2} \qquad (4.16)$$

[$A = 1$]

Equation (4.16) represents the *magnitude of the receiving-end voltage of the reactive power-constrained line during maximum power transfer and when there is no control on the receiving-end bus voltage.*

The amount of power under this condition is obtained on simplification of Eqs. (4.12) and (4.14) as

$$P_{R_{max}} = \frac{V_{max}^2}{B}\cos\phi \qquad (4.17)$$

where $P_{R_{max}}$ is the maximum amount of power transfer for this line.

The magnitude of the receiving-end voltage at voltage stability limit has been determined in Section 4.2(a) while the same at steady state stability limit has been obtained in Section 4.2(b). In a power line, neglecting the effect of loads, both the Eqs. (4.8) and (4.16) are identical.

4.3 EXPRESSION OF MAXIMUM POWER ANGLE AT VOLTAGE STABILITY LIMIT

From Eq. (4.8) it is evident that at the voltage stability limit, the receiving-end voltage of a reactive power-constrained line is given by

$$V_{cri} = \frac{E}{2\cos\delta}$$

while the maximum receiving-end voltage at steady state stability limit is given by Eq. (4.16)

$$V_{max} = E\left[\frac{1}{2(1+\sin\phi)}\right]^{1/2}$$

The problem of voltage stability is closely related to the stability of receiving-end bus while the problem of traditional steady state stability is associated mainly with the generators. However, these two concepts are quite similar and the analytical expressions of the constituent and governing parameters of these two phenomena can be identical as indicated earlier under a set of specific assumptions. The voltage stability limit and steady state stability limit being coincident, the maximum power angle at voltage stability limit can be obtained when

$$\frac{E}{2\cos\delta} = E\left[\frac{1}{2(1+\sin\phi)}\right]^{1/2}$$

or

$$\frac{E^2}{4\cos^2\delta} = \frac{E^2}{2(1+\sin\phi)}$$

or

$$(1+\sin\phi) = 2\cos^2\delta$$

i.e.

$$\delta = \frac{\pi}{4} - \frac{\phi}{2} \quad (4.18)$$

Thus, $\delta_{lim} = \left(\frac{\pi}{4} - \frac{\phi}{2}\right)$ represents the *limiting power angle in the reactive power-constrained line*.

Using the above expressions, in a typical system (having $B = 0.1$ p.u. and $E = 1.00$ p.u.), the maximum receiving-end bus voltage at unity power factor has been obtained as 0.707 p.u. corresponding to a maximum power transfer of 4.998 p.u. at the limiting power angle 45°.

Physically, this result signifies that the receiving-end bus voltage drops to an unacceptable value during maximum power transfer in a typical power line at unity power factor (UPF). The receiving-end bus voltage drops further with lagging loads indicating that large blocks of power transfer in a transmission line is not feasible when the receiving-end bus voltage is uncontrolled. The amount of power transfer has to be kept low such that the receiving-end bus voltage remains within the acceptable limits for power transmission lines with receiving-end voltage uncontrolled.

4.4 RECEIVING-END BUS VOLTAGE IN A WEAK TRANSMISSION LINE UNDER NORMAL OPERATING CONDITION

In the lossless frame, the power flow equations given by Eq. (4.2) and (4.3) can be simplified and represented as

$$P_R = \frac{EV}{X} \sin \delta \qquad (4.19)$$

$$Q_R = \frac{EV}{X} \cos \delta - \frac{V^2}{X} \qquad (4.20)$$

while the relation between P_R and Q_R is given by

$$Q_R = P_R \tan \phi \qquad (4.21)$$

Comparing Eqs. (4.20) and (4.21) and using Eq. (4.19),

$$\left[\frac{EV}{X} \sin \delta\right] \cdot \tan \phi = \frac{EV}{X} \cos \delta - \frac{V^2}{X}$$

which yields

$$V = E\,(\cos \delta - \sin \delta \tan \phi) \qquad (4.22)$$

Equation (4.22) represents *the value of the receiving-end bus voltage under normal operating condition of a radial line in terms of power angle, sending-end voltage and power factor at the load end.*

4.5 EXPRESSION OF LIMITING REACTIVE POWER REQUIREMENT

The determinant of the Jacobian of the load flow equations (given by Eqs. (4.19) and (4.20)) in a lossless frame is given by

$$\Delta[J] = \frac{1}{X}(-E^2 V + 2V^2 E \cos \delta) \qquad (4.23)$$

Applying the same reasoning for assuming the condition of $\Delta[J] = 0$, at voltage stability limit, the receiving-end voltage is given by

$$E = 2V \cdot \cos \delta \qquad (4.24)$$

Utilising Eq. (4.24) in Eq. (4.20), at the limiting stage of voltage stability

$$Q_{\lim} = \frac{2V^2 \cdot \cos^2 \delta}{X} - \frac{V^2}{X}$$

$$Q_{\lim} = \left(\frac{V^2}{X}\right) \cdot \cos 2\delta \qquad (4.25)$$

where, $Q_{\lim}$ denotes the *limiting value of the reactive power transfer in a transmission system at the limiting stage of voltage collapse*. Figure 4.1 represents the profile of calculated limiting value of reactive power transfer at any stable voltage state for a simulated 400 kV 'Bundle Conductor' system, when the system load is maintained constant following the increase in the reactive burden due to stipulated contingency for discrete voltage stages within 10% limit of transmission voltage regulation.

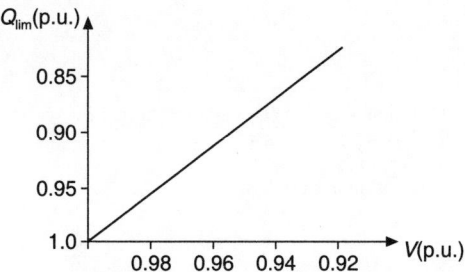

Figure 4.1 Reactive power limit to have voltage stability for different receiving-end voltages.

It simply suggests that for a load voltage range from 1.0 p.u. to 0.9 p.u., the stable voltage state will be maintained provided the system has the corresponding limiting values of reactive power transfer capability as given by Eq. (4.25). Reactive power demand beyond the limit given by this expression corresponding to any voltage, will lead to voltage collapse.

4.6 RELATION BETWEEN REACTIVE POWER VARIATION AND SYSTEM STABILITY

A unique method of estimating the stability limit of a power system using load flow analysis had been first suggested by Venikov et al. For the assessment of the stability limit and the corresponding stability margin, successive changes in the initial operating conditions can be made and the stability of each changed condition can then be verified.

The changes in the operating conditions can be obtained by any one or by a combination of the following methods:

(a) By increasing the load and generation at certain nodes of the network.
(b) By redistribution of generations.
(c) By creating the deficiency of reactive volt-amperes in a part of the system.

The stability of the changed system can be verified by noting down the sign of the determinant of the Jacobian obtained from a load flow analysis (as suggested by Venikov). It has also been suggested that the system remains in a stable state provided the sign of the determinant for the new operating point is the same as that of the initial operating point. Instability occurs if this condition is not satisfied.

Using the above criteria, the effect of reactive power variation on the system stability can be studied. The simulation has been carried for a standard Ward & Hale 6 Bus Power System with Bus 1 as the slack bus, Bus 2 the PV bus and the remaining buses being the load or PQ buses.

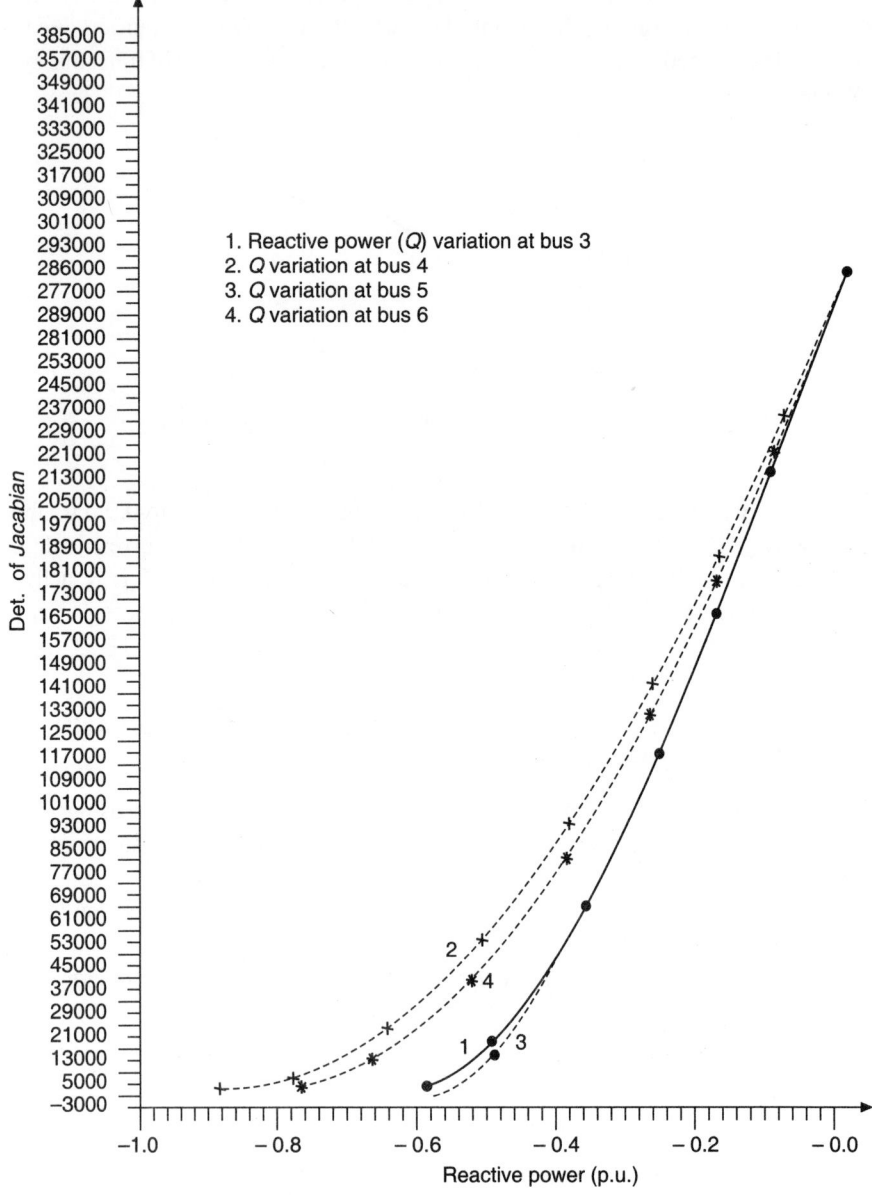

Figure 4.2 Reactive power vs Det. of Jacobian.

At each load bus, inductive reactive power is changed in steps until the determinant changes its sign indicating the change in the status of system stability. The profiles of the Jacobian for the variation of the reactive power injection at load buses have been shown in Figure 4.2.

It can be observed that the system stability is lost if the reactive power status at bus 5 is more than 0.6 p.u. However, an increase in the reactive power change at bus 4 can be allowed up to 0.9 p.u. before the system becomes unstable. This simulation exhibits the capability of the reactive power at the bus in maintaining system stability. This simulation also indicates the need for installation of the reactive power compensators at the selective load buses such that the comprehensive stability of the model system can be enhanced.

4.7 REACTIVE POWER SENSITIVITY GOVERNED VOLTAGE STABILITY

From the conventional power flow equations, using the load flow analysis models, the nodal difference equation for bus powers in matrix form is given by

$$\begin{bmatrix} \Delta P \\ \Delta Q \end{bmatrix} = \begin{bmatrix} \frac{\partial P}{\partial \delta} & \frac{\partial P}{\partial V} \\ \frac{\partial Q}{\partial \delta} & \frac{\partial Q}{\partial V} \end{bmatrix} \begin{bmatrix} \Delta \delta \\ \Delta V \end{bmatrix} \quad (4.26)$$

With the analogous reasoning of the FDLF (Fast Decoupled Load Flow) model, it is considered that

$$\frac{\partial P}{\partial V} = 0 \text{ and } \frac{\partial Q}{\partial \delta} = 0 \quad (4.27)$$

Then
$$[\Delta Q] = \left[\frac{\partial Q}{\partial V}\right][\Delta V] \quad (4.28)$$

For small changes in ΔQ and ΔV, the diagonal elements $(\delta Q_i/\delta V_i)$ represent the steady state stability indices.

Hence, if the generator buses are to be eliminated,

$$[\Delta Q_L] = \left[\frac{\partial Q_L}{\partial V_L}\right][\Delta V_L] \quad (4.29)$$

Table 4.1 represents the values of the diagonal elements of the Ward-Hale system.

Table 4.1 Magnitude of diagonal elements $(\delta Q_i/\delta V_i)$

Bus No.	3	4	5	6
Magnitude of diagonal element	6.90	10.65	4.00	6.80

An insight into this analysis reveals that the change in the magnitude of $(\delta Q_i/\delta V_i)$ of the load bus through "Zero" indicates the instability, i.e. the system (or the load bus) remains stable, provided (dQ_i/dV_i) is positive. Its zero value indicates the limiting stage of stability of negative value, i.e. the instability. Then, observing the data of Table 4.1, it is easy to conclude that Bus 4 is the most stable while Bus 5 is the least stable. This also coincides with the results of Venikov regarding system stability. The order of stability of the buses has now been shown in Table 4.2 for the assumed system.

Table 4.2 Order of stability

Bus No.	Order/status of stability			
4	+	+	+	+
3	+	+	+	
6		+	+	
5		+		

If L denotes the index of steady state stability considering the reactive power sensitivity to voltage, the criterion of system stability can be represented by

$$\left.\begin{array}{l} L > 0 \text{ when the system is stable,} \\ L = 0 \text{ at the limiting stage of stability,} \\ L < 0 \text{ at the unstable state.} \end{array}\right\} \quad (4.30)$$

Also, the steady state stability margin (SM), for the reactive power change in any load bus, can be expressed as

$$\text{SM} = L_{\text{operating}} - L_0 \quad (4.31)$$

where $L_{\text{operating}}$ denotes the value of $(\delta Q_i/\delta V_i)$ during any steady state operation and L_0 the value of the same at steady state stability limit.

Obviously the value of 'SM' in this context differs from the conventional steady state stability margin. The concept of margin here indicates the stability margin of the system with respect to the operation of the system for maximum reactive power variation, while the conventional stability margin represents the margin of stability of maximum real power variations in systems. The magnitude of 'SM' in the present analysis can be enhanced utilising the reactive power compensators at the load buses, thus insisting the system to be more stable.

4.8 VOLTAGE STABILITY AND LOAD FLOW SOLUTION

Next, taking into account the same example, the reactive power status is further changed in Bus 6 such that the power factor of that bus becomes more and more lagging ($Q > 0$ denotes the lagging power factor load). Figures 4.3 and 4.4 exhibit the profile of V_6 against Q_6 and $|J|$ against Q_6 [Voltage and magnitude of Jacobian vs reactive power demand at Bus 6 respectively].

It is evident that the number of the solutions change with the value of Q_6. Point D in Figure 4.3 and point E in Figure 4.4 denote the voltage stability limit where the Jacobian is singular. In other words, as the reactive power load in lagging power factor increases, simulating the heavy load condition, the load flow solution provides a unique root. Till that point is reached, two roots of the load flow equations exist. The system operation corresponding to the roots in the higher voltage state indicates the feasible solution in case the system is heavily loaded. The operation in the zone corresponding to the lower roots causes the system to draw huge current from the source resulting in more and more series reactive loss which in turn depresses the voltage further in the transmission line. This effect is cumulative and may lead to voltage collapse.

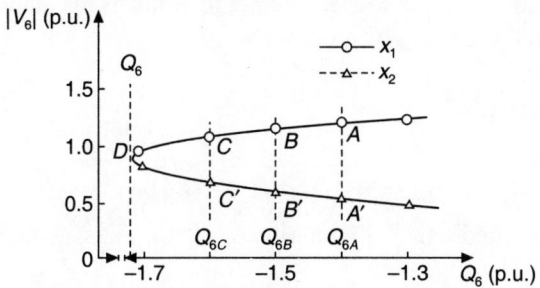

Figure 4.3 Relationship between the VAR injection Q_6 and voltage magnitude $|V_6|$ ($\omega C_6 = 1.96$, see Table 4.1).

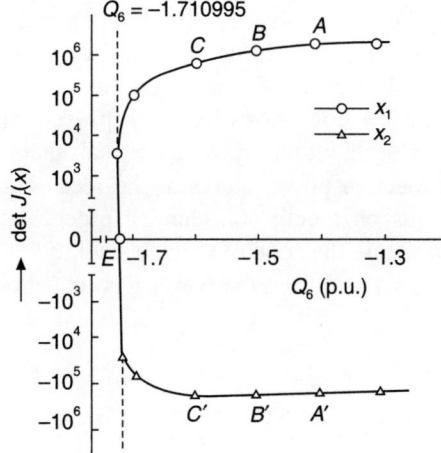

Figure 4.4 Relationship between the VAR injection Q_6 and det J_r ($\omega C_6 = 1.96$).

Thus, multiple load flow solutions exist till the system reaches its limiting stage of voltage stability. The higher voltage magnitudes of the roots of the solution represent feasible operation under medium loading. With increasingly lagging loads, the system becomes heavily loaded and at a particular reactive power status the system reaches its limiting stage of voltage stability (this state has earlier been shown to be under the condition when the Jacobian of the load flow equation is singular).

4.9 LOADING LIMIT OF A TRANSMISSION SYSTEM AT VOLTAGE STABILITY

Here an analytical expression to specify the reaction of loading of an EHV radial system on voltage stability is shown. The performance of a power system can be represented by the load flow analysis where the Jacobian is conventionally represented by Eq. (4.6) in a lossless frame, in terms of real and reactive power status at the load bus and is given by

$$[J] = \begin{bmatrix} Q + (V^2/x) & (P/V) \\ (-P) & -2(V/x) + (Q/V) + (V/x) \end{bmatrix} \quad (4.32)$$

Here
$$\Delta[J] = [V^4 - x^2 (P^2 + Q^2)]/Vx^2 \quad (4.33)$$

Applying the condition of singularity, $\Delta|J| = 0$. Thus, at the limit of voltage stability, Eq. (4.33) is given by

$$V_{cri}^4 - x_{cri}^2 (P^2 + Q^2) = 0$$

i.e.
$$(P^2 + Q^2) = V_{cri}^4 / x_{cri}^2$$

i.e.
$$|S| = V_{cri}^2 / x_{cri} \quad (4.34)$$

where, V_{cri} is given by Eq. (4.8) and x_{cri} is the system reactance of critical state.

Equation (4.34) represents the basic analytical expression for determining the loading limit of an EHV power line at its limiting stage of voltage stability. However, this expression does not directly represent the reactive power contribution. A deeper insight into this problem reveals that the effect of reactive power capability on reactive power status at the receiving-end bus is a very important factor, and must be directly considered in order to assess the real picture of power flow pattern within the zone of voltage stability.

Again, the elimination of δ from the power flow equations

$$P = \frac{EV}{x} \sin \delta \quad (4.35)$$

and
$$Q = \frac{EV}{x} \cos \delta - \frac{V^2}{x} \quad (4.36)$$

results in the system voltage equation and is represented by Eq. (4.37).

$$V^4 + V^2 (2Qx - E^2) + x^2 (P^2 + Q^2) = 0 = \psi(P, Q, V) \quad (4.37)$$

Analysis of the system voltage equation (Eq. 4.37), which is the function of voltage as well as real and reactive powers, about the point of voltage collapse (i.e. when $d\psi/dV = 0$), with elimination of V evolves

$$P^2 x^2 + x E^2 Q = \frac{E^4}{4} \quad (4.38)$$

This expression represents the equation of a parabola, if plotted in a P-Q (two dimensional) frame. This expression directly reveals the loading limit of a radial line for any value of the reactive power status at the load bus.

For a typical power line, Eq. (4.37) can be geometrically plotted in a three-dimensional plane with P, Q and V as axes (Figure 4.5). The 3D contour represents the pictorial view of the voltage stability contour where the 'bold line' (i.e. the fold of the pictorial figure in the 3D plane) represents the voltage stability limits of the power system. It may be observed that this fold is intercepted by another group of different lines which represent different reactive power status at the load bus. Each point of intersection represents the loading limit of the power line to have voltage stability. With enhancement in the reactive power injection, this loading limit can be further enhanced without producing any risk in the stability of voltage.

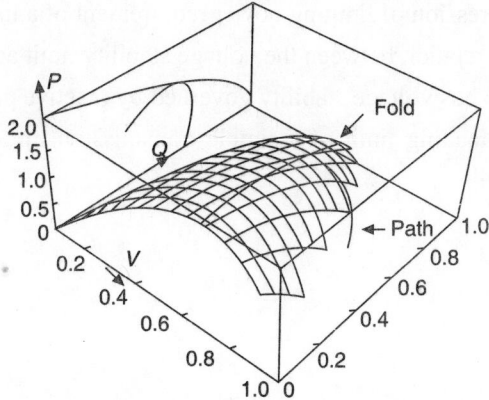

Figure 4.5 3-D view of power system profile.

The 'bold lines' of Figure 4.5 can be projected in the P–Q plane of the 3-D reference frame where the new curve gives another pictorial concept (Figure 4.6) indicating that loading limits are governed by different limits of reactive power condition at the receiving-end bus. The point of intersection of the reactive power characteristic with the fold denotes the bifurcation point representing the loading limit to ensure voltage stability.

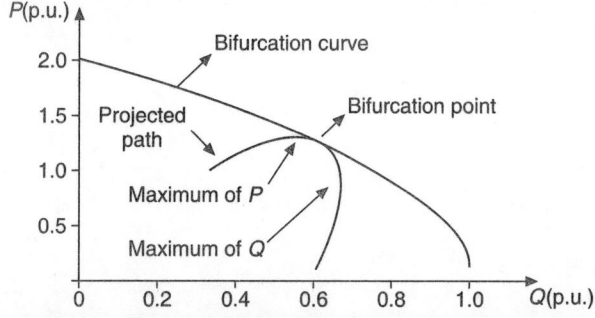

Figure 4.6 A pictorial view of loading limits governed by limits of power conditions.

EXERCISES

1. Develop suitable expressions for the magnitude of the receiving-end voltage at steady state stability limit.
2. What is the expression of the receiving-end voltage at voltage stability limit of a load bus?
3. Find an expression of the power angle at the load bus in a simple line model at voltage stability limit.
4. How do you express the receiving-end bus voltage in terms of power angle and power factor angle?
5. Find an expression of limiting power requirement of a line.
6. Discuss the relation between the voltage stability limit and the Jacobian.
7. How is load bus voltage stability governed by reactive power sensitivity?
8. Obtain the loading limit of a simple transmission line model at voltage stability limit.

Chapter 5

Voltage Stability Indicators

5.1 INTRODUCTION

The topic of voltage stability has been investigated since the last 25 to 30 years across the globe and numerous works have been resported in developing the indicators of voltage stability. Voltage stability primarily depends on reactive power status of the load bus along with a few other governing parameters, e.g. system reactance, type of load etc.; hence it is necessary to include a discussion on indicators of voltage stability. In the following sections a few indicators of voltage stability have been listed though the list is not exhaustive. Importance is given to selection of those indicators which have practical application and have been used extensively in different areas.

5.2 A FUNDAMENTAL INDICATOR USING P-V AND Q-V CURVES

Early attempts to investigate voltage unstable conditions were based on attempts to improve the static load flow solution for a heavily loaded system. But at the voltage collapse point, there is no real steady state solution of the load flow. Later, dual solutions (with two different voltages for the same power delivered) were observed. For such cases, there was a single point beyond which it became impossible to solve the power flow.

Voltage collapse proximity indicators generally used the distance of the two solution points as the indicator of proximity to collapse since this distance decreases as the point of maximum loadability approaches.

Figure 5.1 shows the *P-V* diagram of a typical system at a particular operating point. For a particular operating power, there are two solution points. The upper point V_U is the desired operating point since it has a higher voltage value. It can be seen that the distance between the two solution points ΔV tends to zero as the margin of power P_m between the operating point and the point of maximum power approaches 0.

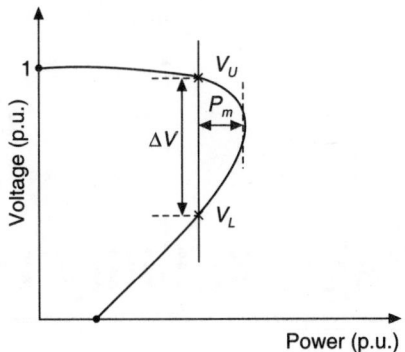

Figure 5.1 P–V diagram of a particular system.

However, P–V curves do not take into account the reactive power component of the load. To include the reactive component, a third dimension must be added.

Figure 5.2 shows the trajectory of the load increase when active and reactive power change arbitrarily. The voltage stability boundary is represented by a projection on the P–Q plane (bold curve). From this curve, it can be observed that:

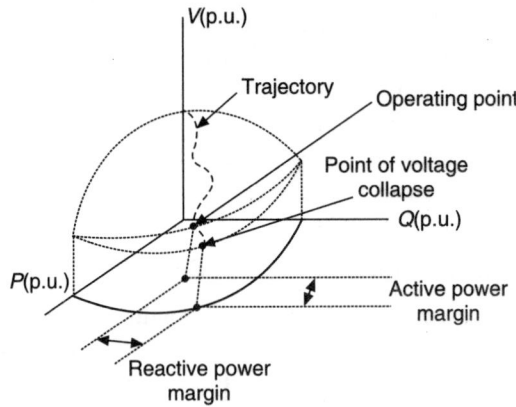

Figure 5.2 Trajectory of load.

1. There are many possible trajectories to voltage collapse.
2. Active and reactive power margins depend on the initial operating point and the trajectory of collapse. Hence there are many voltage collapse proximity indicators. They are usually based on the measurement of the state of a given system under stress and derivation of certain parameters which indicate the closeness to instability of that system. It may be better to use a directly measurable parameter in real time domain to indicate proximity to collapse. An example of such an indicator is the sensitivity of

the generated reactive powers with respect to the load parameters. When the system is close to collapse, small increases in load will cause large increases in reactive power absorption in the system. This increased demand of reactive power must be supplied by dynamic sources of reactive power in the region. At the point of collapse, the rate of change of generated reactive power with respect to load increase at the bus tends to infinity.

Other useful indicators are power margins—margin of active power or margin of reactive power. Figure 5.3 shows the Q–V diagram of a bus in a particular power system at four different loads: P_1, P_2, P_3, P_4. The Q-axis shows the amount of additional reactive power that must be injected into the bus to operate at a given voltage. The operating point is the intersection of the power curve with the voltage axis, where no reactive power is required to be injected or absorbed. If the slope of the curve at the intersection point is positive, the system is stable, because any additional reactive power will raise the voltage and vice-versa.

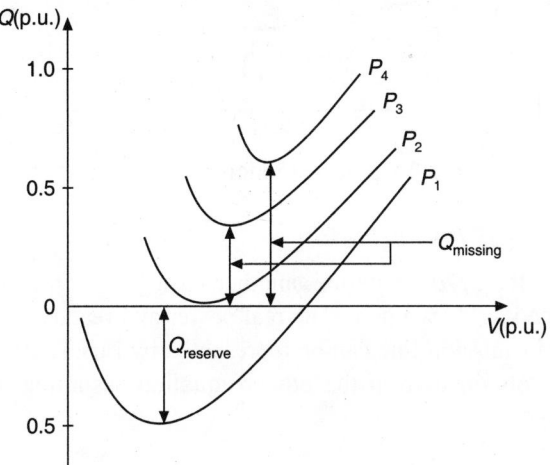

Figure 5.3 Q–V Curve.

Hence for P_1 load, there is a reserve of reactive power that can be used to maintain stability even if the load increases. For P_2 load the system is marginally stable. For higher loads P_3 and P_4 the system is not stable (since a certain amount of reactive power must be injected into the bus to cause an intersection with the voltage axis). Thus the measure of Q reserve gives an indication of the margin between stability and instability.

5.3 CRITERIA OF VOLTAGE STABILITY

5.3.1 *dE/dV* Criterion

The bus voltage at the receiving end in a radial system supplied by a constant voltage source E can be represented as

$$V = \sqrt{\left(\frac{E^2 - Q \cdot x}{E}\right)^2 + \left(\frac{P \cdot x}{E}\right)^2} \qquad (5.1)$$

The profile of E vs. V is plotted in Figure 5.4 from which it is evident that the critical state of voltage stability at the receiving-end bus can be represented by the criterion $(dE/dV) = 0$, at any constant flow of power. Stability is maintained provided $dE/dV > 0$.

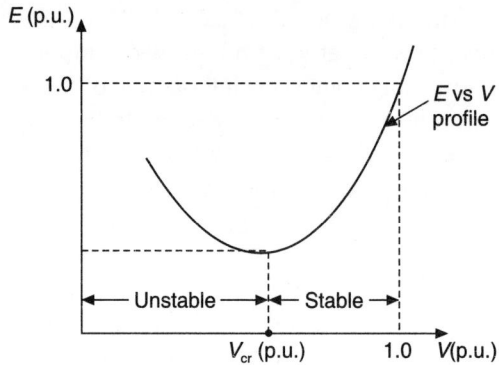

Figure 5.4 dE/dV criterion in voltage stability.

Application of (dE/dV) criterion

While applying the dE/dV criterion, another basic criterion of voltage stability can be developed as shown here. The real power and reactive power status in a basic power transmission line can be represented by Eqs. (5.2) and (5.3) which can readily be obtained from the power equation assuming $A = 1 \angle \alpha°$ and $\beta = \angle(90° - \alpha)$.

$$P = \frac{EV}{Z} \sin(\delta + \alpha) - \frac{V^2}{Z} \sin \alpha = f_1(E, V, \delta) \qquad (5.2)$$

$$Q = \frac{EV}{Z} \cos(\delta + \alpha) - \frac{V^2}{Z} \cos \alpha = f_2(E, V, \delta) \qquad (5.3)$$

Considering the voltage dependent loads in the load bus, the real and reactive power demands are given by

$$P_d = f_3(V) \qquad (5.4)$$

and
$$Q_d = f_4(V) \qquad (5.5)$$

also, in the assumed lossless system,

and
$$\left.\begin{array}{l} P = P_d \\ Q = Q_d \end{array}\right\} \qquad (5.6)$$

For small increments in the system variables and to have a stable voltage state, $dE/dV > 0$ from which it can be shown that

$$\begin{vmatrix} \dfrac{\partial P}{\partial \delta} & \left(\dfrac{\partial P_d}{\partial V} - \dfrac{\partial P}{\partial V}\right) \\ \dfrac{\partial Q}{\partial \delta} & \left(\dfrac{\partial Q_d}{\partial V} - \dfrac{\partial Q}{\partial V}\right) \end{vmatrix} > 0$$

Substituting the value of differentials of P and Q with respect to δ and V in the above equation, the result yields

$$\cos(\delta + \alpha)\left(\dfrac{\partial Q_d}{\partial V} + \dfrac{2V}{Z}\cos\alpha\right) + \sin(\delta + \alpha)\left(\dfrac{\partial P_d}{\partial V} + \dfrac{2V}{Z}\sin\alpha\right) - \dfrac{E}{Z} > 0 \quad (5.7)$$

which represents the necessary condition of voltage stability.

In an EHV power system, the (r/x) ratio is very small, $\alpha = 0$ as $\beta \approx 90°$ and the system is purely reactive. Hence Eq. (5.7), on simplification, becomes

$$\cos\delta\left(\dfrac{dQ_d}{dV} + \dfrac{2V}{x}\right) + \sin\delta\left(\dfrac{\partial P_d}{\partial V}\right) - \left(\dfrac{E}{x}\right) > 0 \quad (5.8)$$

i.e.
$$\left(\dfrac{dQ_d}{dV}\right) > \dfrac{E}{x\cos\delta} - \left(\dfrac{2V}{x} + \dfrac{\partial P}{\partial V}\cdot\tan\delta\right) \quad (5.9)$$

Considering the monopoly of induction motors and decoupling effect at the load bus, Eq. (5.9) modifies to

$$\left(\dfrac{dQ_d}{dV}\right) > \dfrac{E}{x\cos\delta} - \dfrac{2V}{x} \quad (5.10)$$

in the voltage stable state.

Thus, within the zone of voltage stability,

$$\left(\dfrac{dQ_d}{dV}\right) > \dfrac{E}{x\cos\delta} - \dfrac{2V}{x} \quad (5.11)$$

and at the voltage stability limit,

$$\left(\dfrac{dQ_d}{dV}\right) = \dfrac{E}{x\cos\delta} - \dfrac{2V}{x} \quad (5.12)$$

Beyond the zone of voltage stability,

$$\left(\dfrac{dQ_d}{dV}\right) < \dfrac{E}{x\cos\delta} - \dfrac{2V}{x} \quad (5.13)$$

Here, P_d and Q_d can be represented by polynomials in V and are given by

$$P_d = \sum_{m=0}^{n} a'_m V^m$$

giving

$$\frac{\partial P_d}{\partial V} = \sum_{m=0}^{n} a'_m \cdot m \cdot V^{m-1}$$

and

$$Q_d = \sum_{m=0}^{n} a^n_m \cdot m$$

giving

$$\frac{\partial Q_d}{\partial V} = \sum_{m=0}^{n} a^n_m \cdot m \cdot V^{m-1}$$

Again, representing the left hand side of Eq. (5.8) given by

$$\left[\cos\delta \left\{ \left(\frac{\partial Q_d}{\partial V} \right) + \left(\frac{2V}{x} \right) \right\} + \sin\delta \left(\frac{\delta P_d}{\delta V} \right) - \frac{E}{x} \right]$$ as L, an indicator of voltage stability has been obtained for any load bus in a simple power system whence in a voltage stable state $L > 0$. The passage of L from positive to negative values through its zero value indicates the excursion of voltage stable zones from stability to instability.

A simulation of the theory presented for a typical 250 km power system link with ACSR bundle conductor double circuit, 400 kV transmission line on pyramid type steel tower, reveals that the system voltage stability can only be maintained provided the link impedance is low (Figure 5.5). The change in the link impedance has been simulated by contingency of the line or the terminal equipment. A system enters into a voltage unstable zone with change in system reactance which also affects the receiving-end voltage magnitude. Thus (dE/dV) criterion provides a useful tool to anticipate the voltage stable states in the system for contingencies in a power line.

At voltage stability limit, Eq. (5.10) may be presented as

$$\frac{E}{x} = \cos\delta \left(\frac{dQ_d}{dV} + \frac{2V}{x} \right)$$

or

$$I_{sc} = \cos\delta \left(\frac{dQ_d}{dV} + \frac{2V}{x} \right) \quad \text{as } E = (I_{sc} \cdot x)$$

i.e. to have voltage stability, from the preceding analysis,

$$\cos\delta \left(\frac{dQ_d}{dV} + \frac{2V}{x} \right) \geq I_{sc}$$

or

$$E \cos\delta \left(\frac{dQ_d}{dV} + \frac{2V}{x} \right) \geq E I_{sc} \qquad (5.14)$$

Figure 5.5 represents the profile of the source short-circuit capacity against the terminal voltages for different link impedances. It has been observed that the

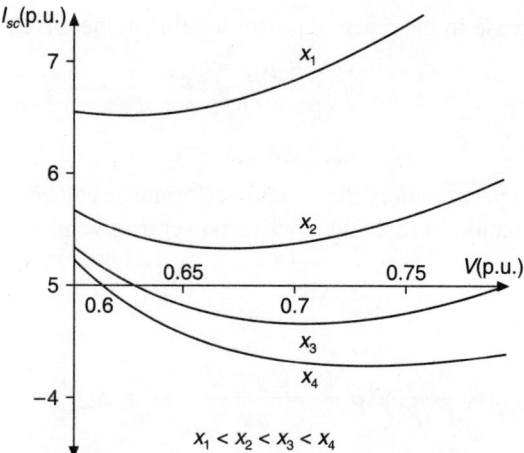

Figure 5.5 Profile of short-circuit strength vs receiving-end voltage of a radial power system at different system reactances.

system reactance being high, the system enters into the zone of voltage instability with change in the value of I_{sc}, i.e. when the contingency occurs. However, with lower values of the system reactance the voltage stability of the system improves.

5.3.2 dQ/dV Criterion

It can been shown that assuming constant frequency of operation and constant terminal voltages, at any constant power input from the turbine, the critical condition of steady state stability occurs when $\partial P/\partial \delta = 0$.

However, assuming the active and reactive power balance at the load node, if both the generators at the two ends of a longitudinal tie line are assumed to be identical (Figure 5.6) and being replaced by a single generator characteristic with

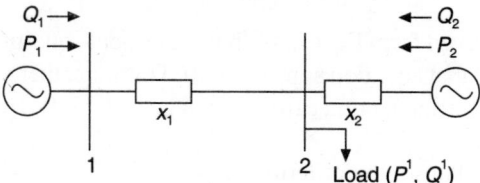

Figure 5.6 A power line comprising of two generators at the two ends.

P_g and Q_g real and reactive power generation, and if $P_1 = P_2 = P$, $Q_1 = Q_2 = Q$, $\Delta\delta_1 = \Delta\delta_2 = \Delta\delta$, then

$$P_1 + P_2 = P_g \tag{5.15}$$
$$Q_1 + Q_2 = Q_{gl} \tag{5.16}$$

where Q_{gl} is the total load reactive power.

For any increase in the reactive power loading alone, given by ΔQ, let
$$Q_{gl} + \Delta Q = Q_g \tag{5.17}$$
Also,
$$\Delta P_1 = \Delta P_2 = 0$$
$$\Delta \delta = 0 \tag{5.18}$$

For the small perturbation, the system performance can be represented by the state variable equation of real and reactive power as given by

$$\frac{\partial P_g}{\partial \delta} \Delta \delta + \frac{\partial P_g}{\partial V} \Delta V = 0$$

and

$$\frac{\partial Q_{gl}}{\partial \delta} \Delta \delta + \frac{\partial (Q_{gl} - Q_l)}{\partial V} \Delta V = \Delta Q_l$$

The simplification of the above two equations yields

$$\Delta V = \frac{\Delta Q_l}{\dfrac{\partial (Q_{gl} - Q_l)}{\partial V} - \dfrac{\dfrac{\partial P_g}{\partial V} \times \dfrac{\partial Q_{gl}}{\partial \delta}}{\dfrac{\partial P_g}{\partial \delta}}} \tag{5.19}$$

Assuming the concept of decoupling $(\partial P_g / \partial V) = 0$, hence Eq. (5.19) becomes

$$\Delta V = \frac{\Delta Q_l}{\dfrac{\partial (Q_{gl} - Q_l)}{\partial V}}$$

$$= \frac{\Delta Q_l}{\dfrac{\partial Q_{gl}}{\partial V} - \dfrac{\partial Q_l}{\partial V}} \tag{5.20}$$

It may be observed form Eq. (5.20) that the critical condition will occur when the denominator of that equation approaches 0. This time, there will be unbounded voltage change at the nodal point causing voltage instability. Hence at the limiting stage of voltage collapse, the criterion is $\dfrac{\partial (Q_{gl} - Q_l)}{\partial V} \simeq \dfrac{d \Delta Q_{gl}}{dV} = 0$ (when the capacity of the system is comparable with that of the load).

The profile of reactive power generation and demand (Figure 5.7), from the basic power flow equation can be obtained as

$$Q_{gl} = \sqrt{\left(\frac{VE}{X}\right)^2 - P^2} - \frac{V^2}{X} \tag{5.21}$$

while

$$Q_l = \frac{V^2}{X} - \sqrt{\left(\frac{VE}{X}\right)^2 - P^2} \tag{5.22}$$

Q_{gl1}, at different values of the sending-end voltage can be plotted as in Figure 5.7, where $Q_{gl1}, ..., Q_{gl4}$ represent different values of reactive power generations corresponding to $E_1, ..., E_4$, the sending-end bus voltages (here $E_1 > E_2 > E_3 > E_4$).

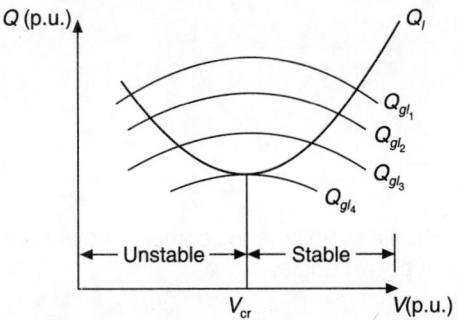

Figure 5.7 Characteristics of Q_l and Q_{gl} to indicate voltage instability.

The point of intersection of these curves corresponds to the steady state condition in the system, while the performance of stability is governed by the difference $\Delta Q = Q_{gl} - Q_l$.

It may be observed from Figure 5.7 that during the normal stable operation, $V > V_{cr}$.

5.4 LOAD VOLTAGE INDICATOR FOR VOLTAGE STABILITY

Assuming the simplest model of a transmission line, the receiving-end current (I_R) is given by

$$I_R = \frac{E-V}{Z} \tag{5.23}$$

Substituting the value of I_R in the fundamental power equation and assuming V to be the reference phasor, simplication yields

$$V^2 - VE + S*Z = 0 \tag{5.24}$$

This is a quadratic equation in V and the graph of V vs. S is given in Figure 5.8. The slope of the curve is given by dV/dS and can be found by taking the first derivative of Eq. (5.24) with respect to S.

i.e.

$$2V\frac{dV}{dS} - E\frac{dV}{dS} + Z = 0$$

or

$$\frac{dV}{dS} = \frac{Z}{E - 2V} \tag{5.25}$$

It is evident form Figure 5.8 that voltage stability is maintained till the slope dV/dS is negative. The limiting condition arrives when $(dV/dS) = 0$ and the system lands to an unstable voltage state as soon as dV/dS is positive. In other words, it means that

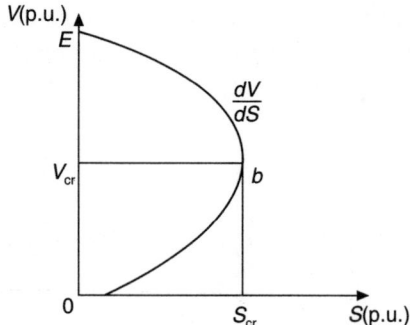

Figure 5.8(a) Profile of voltage vs. complex power neglecting the role of power angle.

1. if $E > 2V$, i.e. $V < E/2$, voltage stability is lost.
2. if $V = E/2$, voltage stability limit is attained.
3. if $V > E/2$, voltage stability is maintained.

The corresponding value of the critical power magnitude at voltage stability limit can be achieved by substituting $V = E/2$ (i.e. the magnitude of the receiving-end voltage at critical limit) in Eq. (5.24) and solving for S. The critical power limit can be found to be equal to $E^2/4Z^*$.

Assuming $E = 1.00$ p.u., Eq. (5.25) can be rewritten as

$$\frac{dV}{dS} = \frac{(Z/2)}{0.5 - V} \tag{5.26}$$

from which, it is evident that for the voltage stable condition, $V > 0.5$ p.u. (or $E/2$) when Z, the system impedance remains constant. Thus, Eq. (5.25) or Eq. (5.26) represents another indicator of voltage stability. It may be noted, at this point, that with any change in Z, the maximum amount of power transfer changes and Figure 5.8 represents the V–S characteristics of a system with different values of line impedance. It may be observed that while V_{cr} remaining the same, the value of S_{cr} decreases with increase in Z. It may be further noted that the above indicator, though very effective in physical understanding, is further simplified assuming the receiving-end voltage to be a reference phasor with its angle to be zero. Without applying any such assumption, utilisation of Eq. (5.23) in fundamental power equation results in

$$VV^* - V^*E + S^*Z = 0 \tag{5.27}$$

Equation (5.27) is not quadratic. The profile of V against S cannot be easily drawn as both are complex numbers. However, at any constant power factor, magnitudes of V against S may be drawn in a plane. The graph will not be a parabola and will assume the shape shown in Figure 5.8.

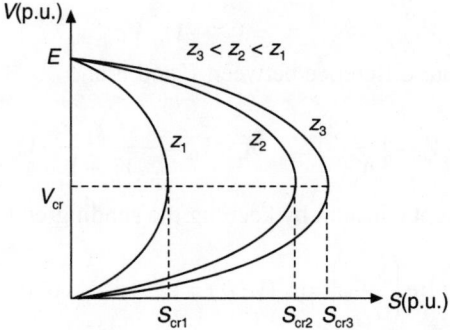

Figure 5.8(b) Voltage vs. complex power characteristics at different line impedances.

5.5 A DIRECT INDICATOR OF VOLTAGE STABILITY AND ITS IMPLICATION ON VOLTAGE STABILITY MARGIN

In this section, an indicator of voltage stability has been established and the condition for voltage collapse has been obtained. The concept of voltage stability margin has been introduced and its effect has been compared with uncompensated and shunt capacitor compensated line(s).

Considering a generation bus connected to a load bus through a transmission line having series admittance Y_L and shunt admittance Y_S, all expressed in per unit quantities the system performance equations may be presented as

$$[I] = [Y][V] \tag{5.28}$$

where
$$[I] = \begin{bmatrix} I_R \\ I_S \end{bmatrix} \quad \text{and} \quad [V] = \begin{bmatrix} V \\ E \end{bmatrix}$$

and the system admittance matrix

$$[Y] = \begin{bmatrix} Y_{LL} & Y_{LS} \\ Y_{SL} & Y_{SS} \end{bmatrix} \begin{bmatrix} (Y_L + Y_S) & (-Y_L) \\ (-Y_L) & (Y_L + Y_S) \end{bmatrix}$$

where V, I_R, S_R and E, I_S, S_S are complex voltages, currents and powers at the receiving (load) bus and sending bus respectively.

The load current I_R is given by

$$I_R = \left(\frac{S_R}{V}\right)^* \tag{5.29}$$

and also from Eq. (5.28)

$$I_R = Y_{LL} V + Y_{LS} E \tag{5.30}$$

A comparison between Eqs. (5.29) and (5.30) yields, after rearrangement,

$$\left(\frac{S_R^*}{Y_{LL}}\right) = V^2 + V_0 \cdot V^* \tag{5.31}$$

$$= V^2 + V_0 \cdot V e^{j\delta} \tag{5.32}$$

where δ is the angular difference between V and E and

$$V_0 = \left(\frac{Y_{LL}}{Y_{LS}}\right) E \quad \text{or} \quad V_0 = -\frac{Y_L}{Y_L + Y_S} \cdot V_S \tag{5.33}$$

Nothing that V_0 is kept constant by keeping the sending-end voltage constant,

$$\text{Re}\left(\frac{S_R^*}{Y_{LL}}\right) = f_1(V,\delta) = V^2 + V_0 V \cos\delta \tag{5.34}$$

and

$$\text{Im}\left(\frac{S_R^*}{Y_{LL}}\right) = f_2(V,\delta) = V_0 V \sin\delta \tag{5.35}$$

The Jacobian corresponding to these power flow equations is then given by

$$[J] = \begin{bmatrix} 2V + V_0 \cos\delta & -V_0 V \sin\delta \\ V_0 \sin\delta & V_0 V \cos\delta \end{bmatrix}$$

where

$$\Delta[J] = [+V_0^2 V \sin^2\delta + V_0 V \cos\delta (2V + V_0 \cos\delta)]$$

$$= +V_0^2 V + 2V_0 V^2 \cos\delta \tag{5.36}$$

Assuming the singularity of Jacobian, at voltage stability limit, $\Delta[J] = 0$, i.e.

$$V_0^2 V + 2V_0 V^2 \cos\delta = 0 \tag{5.37}$$

Equation (5.37) determines the voltage stability limit where the point of voltage instability coincides with the singularity of the Jacobian matrix of the load flow.

Hence from Eq. (5.37),

$$\frac{V}{V_0} \cos\delta = -\left(\frac{1}{2}\right)$$

i.e.

$$\text{Re}\left[\frac{V}{V_0}\right] = -\left(\frac{1}{2}\right) \tag{5.38}$$

Again from Eq. (5.31),

$$S_R^* = V^2 Y_{LL} + V_0 V^* Y_{LL}$$

The above equation can further be simplified by taking the complex conjugate on both sides, and divinding by $(Y_{LL}^* V^2)$ as

$$\frac{S_R}{Y_{LL}^* V^2} - 1 = \frac{V_0^* V Y_{LL}^*}{Y_{LL}^* V^2} = \frac{V_0^*}{V}$$

where

$$\left|1 + \frac{V_0^*}{V}\right| = \left|\frac{S_R}{Y_{LL}^* V^2}\right| \tag{5.39}$$

From Eqs. (5.38) and (5.39),

$$\left|\frac{S_R}{Y_{LL}^* V^2}\right| = 1 \qquad (5.40)$$

Equation (5.40) gives the criterion of voltage stability of the assumed line model. In other words, for a system as described, to retain its voltage stability, the magnitude of the expression $\left[\dfrac{S_R}{Y_{LL}^* V^2}\right]$ must be less than unity. $\left|\dfrac{S_R}{Y_{LL}^* V^2}\right|$ is thus an indicator of voltage stability. From the above analysis the concept of 'Voltage Stability Margin' (M_v) emerges which can be defined as

$$M_v = \left[1 - \left|\frac{S_R}{Y_{LL}^* V^2}\right|\right] \times 100\%.$$

M_v represents the margin of voltage collapse for a system. The value of M_v during maximum power transfer of a system is of interest and thus has been investigated. To establish the validity of the theoretical analysis, a network consisting of a lossless 220 kV, 3-phase, single circuit bundle conductor 200 km transmission line with ACSR 240/40 mm² conductor in the horizontal configuration is simulated as an equivalent π-network having the following line constants.

$$A = D = 0.97598$$
$$B = j\,64.677 \text{ ohm}$$
$$C = j\,7.34 \times 10^{-4} \text{ mho}$$

The shunt branch admittance and the series reactance of the equivalent network are obtained on calculation as

$$Y_S = j\,3.7138 \times 10^{-4} \text{ mho } (X_C = -j\,2692.5 \text{ ohm})$$

and

$$X_L = j\,64.677 \text{ ohm respectively}$$

The profile of the receiving-end voltage vs. the receiving-end complex power in p.u. (both being magnitudes) is shown in Figure 5.9. The lowermost characteristic in this figure is obtained without any shunt capacitor support at load bus while the rest of the characteristics are with different shunt capacitors at load end. The magnitude of $\left[\dfrac{S_R}{Y_{LL}^* V^2}\right]$ is unity at the maximum power transfer for the characteristics, whereas $\left|\dfrac{S_R}{Y_{LL}^* V^2}\right|$ is less than unity above the maximum power transfer line and more than unity below the line. This indicates that the voltage stability margin M_v exists above the maximum power transfer line, it is nil at the maximum power transfer line and vanishes below that line.

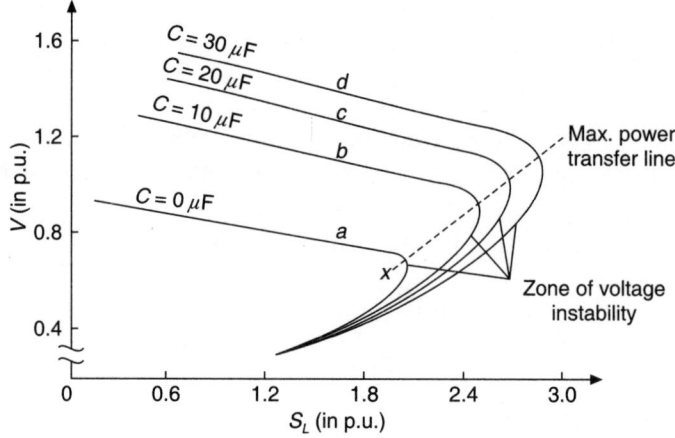

Figure 5.9 Profile of receiving-end voltage with and without shunt capacitive support for various values of the receiving-end real power of an LPS line.

5.6 A VOLTAGE STABILITY INDICATOR BASED ON GOVERNING EFFECT OF LOAD INCREASE

Let the initial load level be represented by P_l^0 and Q_l^0. If γ be the factor simulating the load level, assuming a change in load level to P_l^f and Q_l^f, the power flows can be represented as

$$\Delta P_l = P_l^f - P_l^0 \tag{5.41}$$

and
$$\Delta Q_l = Q_l^f - Q_l^0$$

Also,

$$\begin{bmatrix} \Delta P_l \\ \Delta Q_l \end{bmatrix} = \gamma \begin{bmatrix} P_l^0 \\ Q_l^0 \end{bmatrix} \tag{5.42}$$

The system performance is given by the well-known load flow expression

$$\begin{bmatrix} \Delta P \\ \Delta Q \end{bmatrix} = [J] \begin{bmatrix} \Delta \delta \\ \Delta V \end{bmatrix} \tag{5.43}$$

Here J = full Jacobian matrix.

Equation (5.43) can be solved provided the matrix elements are known. (J is to be computed utilising the system state variables, ΔP is generally known for all buses except the slack bus and ΔQ is known for all load buses.)

Again the sensitivity coefficients are obtained as

(a) $\begin{bmatrix} \Delta \delta \\ \Delta \gamma \end{bmatrix}^f$ for each bus

(b) $\left[\dfrac{\Delta Q}{\Delta \gamma}\right]^f$ for generator bus

(c) $\left[\dfrac{\Delta V}{\Delta \gamma}\right]^f$ for load buses.

The generator reactive power limit can thus be expressed as

$$\left\{Q^f + \left[\dfrac{\Delta Q}{\Delta \gamma}\right]^f \gamma^f = Q_{\max}\right\} \qquad (5.44)$$

where, γ^f can be determined for $[\Delta Q/\Delta \gamma]$.

Hence, for the change in load level, the revaluation of δ and V can be found as

$$\delta^f = \delta^o + \left[\dfrac{\Delta \theta}{\Delta \gamma}\right]^f \gamma^f \qquad (5.45)$$

and

$$V^f = V + \left[\dfrac{\Delta V}{\Delta \gamma}\right]^f \gamma^f \qquad (5.46)$$

This new set of values defines a new state for which the Jacobian becomes J_K. At this new stage, to have voltage stability, the condition $\delta^f < \delta_{\text{cri}}$ and $V^f < V_{\text{cri}}$ must be met. In other words, it may also be concluded that the voltage stability will be maintained till $\left[\dfrac{\delta V}{\delta P}\right] < 0$ or $\left[\dfrac{\delta V}{\delta Q}\right] < 0$ (This is evident from the P–V and Q–V characteristics as well)

These sensitivity coefficients $\left[\dfrac{\partial V}{\partial P} \text{ and } \dfrac{\partial V}{\partial Q}\right]$ may be obtained from the system whose performance is now given by

$$\begin{bmatrix}\Delta P\\ \Delta Q\end{bmatrix} = [J_K]\begin{bmatrix}\Delta \theta\\ \Delta V\end{bmatrix} \qquad (5.47)$$

Hence, Eqs. (5.45) and (5.46) are an indicator of voltage stability, utilising the load flow analysis and when the system is governed by enhancements in load.

5.7 VOLTAGE STABILITY INDEX (L)

In this section, an index of voltage stability has been established and voltage stability margin has been determined from the voltage stability index (L). Let us consider a system of total n number of buses with $1, 2, ..., g$ generator buses, $g+1, g+2, ..., g+l$ PV buses and $g+l+1 ... n$ the remaining PQ buses. For this system, the simplified representation of L for the jth load bus is derived by neglecting the influences of real part of Y-Line matrix (since x/R ratio is large for overhead transmission lines)

$$L_j = \left| 1 - \sum_{i=1}^{i=g} F_{ji} \frac{V_i}{V_j} \right| \qquad (5.48)$$

where $j = g + l + 1 \ldots n$ and all the terms within Σ on the RHS of Eq. (5.48) are complex quantities. The elements of F_{ji} are calculated from the $[Y]$ bus matrix for the network as follows

$$[F] = -[B'']^{-1}[B_{LG}] \qquad (5.49)$$

where $[B'']$ is the imaginary part of the matrix $[Y_{LL}]$ and $[B_{LG}]$ is the imaginary part of the matrix $[Y_{LG}]$. $[Y_{LL}]$ and $[Y_{LG}]$ are the submatrices of the Y_{BUS} matrix. The index L_j indicates proximity of voltage collapse of a power system, a value of $L_j = 1$ indicates the voltage collapse condition at bus j. Hence, the global indicator L describing the stability of the complete system is given by

$$L = \text{maximum of } L_j \text{ for all } j \text{ (load buses)}$$

The indicator L is a quantitative measure for the estimation of the distance of the actual state of the system to the stability limit. The local indicators L_j permit the determination of those buses from which a collapse may originate. The stability limit will be reached for $L = 1$. The stability margin in this case is obtained as the distance L from a unit value, i.e. $(1 - L)$.

The advantage of this method lies in the simplicity of the numerical calculation and expressiveness of the results.

5.8 SINGULAR VALUE DECOMPOSITION

In power systems, one is usually interested in determining the singularity of the Jacobian associated with the system dynamic equations, as this singularity corresponds to a bifurcation point, an undersirable condition of the system. Generally the static voltage stability analysis is based on the modal analysis of the power flow Jacobian matrix as shown in Eq. (5.50).

$$\begin{bmatrix} \Delta P_{PQ,PV} \\ \Delta Q_{PQ} \end{bmatrix} = \begin{bmatrix} J_{PQ} & J_{PV} \\ J_{QQ} & J_{QV} \end{bmatrix} \begin{bmatrix} \Delta \delta \\ \Delta V_{PQ} \end{bmatrix} \qquad (5.50)$$

where

$\Delta P_{PQ,PV}$ = the incremental change in active power of PQ and PV buses.
ΔQ_{PQ} = the incremental changes in reactive power of PQ buses.
$\Delta \delta, \Delta V_{PQ}$ = the vectors that contain the incremental changes in bus voltage angle and bus voltage magnitude.

The elements of the Jacobian matrix represent the sensitivies between the power flow bus voltage changes. According to the classical static voltage stability analysis, voltage stability is largely affected by the reactive power. So, keeping the active power constant at each operating points, the $Q-V$ analysis can be carried out. Assuming $\Delta P = 0$, we can write

$$\Delta Q_{PQ} = [J_{PQ} J_{QV} - J_{PV} J_{QQ}][\Delta V_{PQ}]$$

$$= [J_{QV} - J_{QQ} J_{PQ}^{-1} J_{PV}] \Delta V_{PQ}$$

$$= [J_R] \cdot \Delta V_{PQ} \qquad (5.51)$$

and
$$\Delta V_{PQ} = J_R^{-1} \Delta Q_{PQ} \qquad (5.52)$$

Based on J_R^{-1}, which is the reduced V–Q Jacobian matrix, the Q–V modal analysis can be carried out. Since J_R^{-1} is a square matrix, the singular value analysis of J_R^{-1} provides the same result as the classical static voltage stability analysis. In general, at saddle mode bifurcation point one has that J_{PQ} is not singular, even though J is singular. In practice, it seems that at bifurcation point

$$\det J_R = \frac{\det J}{\det J_{PQ}} \qquad (5.53)$$

5.9 EXPRESSIONS FOR DIFFERENT INDICATORS TO INVESTIGATE THE VOLTAGE SECURITY OF A POWER SYSTEM

To investigate the voltage security of a weak power system, generally two indicators can be used. They are Fast Voltage Security Index (FVSI) and Line Quality Factor (LQF). Based on the values of these indicators, Line Outage Factor (LOF) and Voltage Security Factor (VSF) can be calculated. The values of LOF and VSF can be used to predict the voltage security of multibus power systems.

Let us consider a two-bus system where the complex power and voltage values of the two buses are known. The impedance of the connecting transmission line between the two buses is $R + jX$, as shown in Figure 5.10.

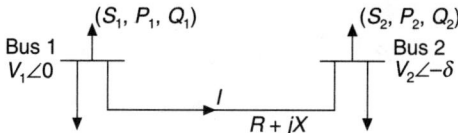

Figure 5.10 Two-bus power system model.

The basic equation of this two-bus system can be written as

$$V_2 \angle -\delta + \left[\frac{P_2 + jQ_2}{V_2 \angle -\delta}\right]^* (R + jX) = V_1 \angle 0 \qquad (5.54)$$

or
$$V_2^2 + (P_2 - jQ_2)(R + jX) = V_1 V_2 \angle \delta$$

or $\quad V_2^2 + P_2R + Q_2X + j(P_2X - Q_2R) = V_2V_1\cos\delta + jV_1V_2\sin\delta$

Now separating the real and the imaginary part of this equation, the following two equations can be obtained.

$$V_2^2 + P_2R + Q_2X = V_2V_1\cos\delta \tag{5.55}$$

$$P_2X - Q_2R = V_1V_2\sin\delta \tag{5.56}$$

Now by substituting the value of P_2 from Eq. (5.56) in Eq. (5.55), we can get

$$V_2^2 + Q_2X + R\left[\frac{V_1V_2\sin\delta}{X} + \frac{Q_2R}{X}\right] = V_1V_2\cos\delta$$

or $\quad V_2^2 + V_1V_2\left(\dfrac{R\sin\delta}{X} - \cos\delta\right) + \left(Q_2X + \dfrac{Q_2R^2}{X}\right) = 0 \tag{5.57}$

For simplicity, we consider that the transmission line resistance has small value. Therefore, Eq. (5.57) can be written as [with $R^2 \to 0$]

$$V_2^2 + V_1V_2\left(\frac{R\sin\delta}{X} - \cos\delta\right) + Q_2X = 0 \tag{5.58}$$

The roots of this equation are

$$V_2 = \frac{-V_1\left(\dfrac{R\sin\delta}{X} - \cos\delta\right) \pm \sqrt{V_1^2\left(\dfrac{R\sin\delta}{X} - \cos\delta\right)^2 - 4Q_2X}}{2}$$

For real solution of V_2, the discriminant should be greater than or equal to zero, i.e.

$$V_1^2\left(\frac{R\sin\delta}{X} - \cos\delta\right)^2 - 4Q_2X \geq 0$$

At critical point

$$V_1^2\left(\frac{R\sin\delta}{X} - \cos\delta\right)^2 = 4Q_2X$$

or $\quad V_1^2\left(\dfrac{\sin\delta\cos\theta}{\sin\theta} - \cos\delta\right)^2 = 4Q_2X$

or $\quad V_1^2\left[\sin(\delta - \theta)\right]^2 = 4Q_2X\sin^2\theta$

∴ FVSI can be defined as

$$\text{FVSI} = \frac{4Q_2X}{V_1^2\left[\sin(\delta - \theta)\right]^2} \tag{5.59}$$

$$\left[\text{as } R \text{ is negligibly small, } \tan\theta = \frac{X}{R} = \infty;\ \theta \approx 90°,\ \text{and } \sin^2\theta \approx 1 \right]$$

So, at critical loading, the value of FVSI should be 1 and at normal loading condition the value of FVSI should be less than 1.

Again from Eqs. (5.55) and (5.56), we can eliminate δ and get

$$V_1^2 V_2^2 = (V_2^2 + P_2 R + Q_2 X)^2 + (P_2 X - Q_2 R)^2 \tag{5.60}$$

By following the assumption that $R = 0$, we get the reduced equation from Eq. (5.60) as

$$V_1^2 V_2^2 = (V_2^2 + Q_2 X)^2 + P_2^2 X^2$$

or $\qquad V_2^4 + V_2^2 (2Q_2 X - V_1^2) + (P_2^2 + Q_2^2) X^2 = 0$

Again, to get the real roots of this equation, the discriminant should be greater than or equal to zero,

i.e. $\qquad (2Q_2 X - V_1^2)^2 - 4(P_2^2 + Q_2^2) X^2 \geq 0$

At the critical point,

$$(2Q_2 X - V_1^2)^2 = 4(P_2^2 + Q_2^2) X^2$$

or $\qquad V_1^4 - 4Q_2 X V_1^2 - 4X^2 P_2^2 = 0$

Therefore the Line Quantity Factor (LQF) can be defined as

$$\text{LQF} = \frac{4X\left(Q_2 + \dfrac{P_2^2 X}{V_1^2}\right)}{V_1^2} \tag{5.61}$$

As the power demand increases, the LQF value of a particular bus approaches 1.

With the help of these indicator values, we can calculate the VSF as follows.

$$\text{VSF} = \frac{\text{Voltage at a particular bus when the indicator reaches its limiting value}}{\text{Voltage of that bus when the indicator is less than the limiting value}}$$

The greater the value of VSF of a particular bus, the more the security of that line.

5.10 VOLTAGE STABILITY EVALUATION BY USING MAXIMUM POWER TRANSFER PHASOR DIAGRAM

From the voltage stability point of view, maximum permissible loading limits should not be exceeded during the operation of the power system. Here, a maximum power transfer phasor diagram is used for easy evaluation of the relations between the major parameters affecting the voltage stability margins.

A power system having any number of generators, transmission lines and loads can be modelled in a Thevenin equivalent form. In Figure 5.11, Thevenin equivalent of power system for load bus k having a load of $S_k = P_k + jQ_k$ is shown.

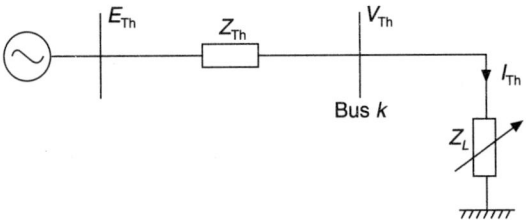

Figure 5.11 Representation of load bus k in a general power system.

From measurements at load bus k, the parameters obtained are

$$V_k = |V_k| \angle 0; \quad I_k = |I_k| \angle \phi \tag{5.62}$$

The Thevenin parameters of power system are

$$E_{Th} = |E_{Th}| \angle \delta; \quad Z_{Th} = |Z_{Th}| \angle \alpha \tag{5.62a}$$

The short-circuit current for bus k is

$$I_{k,sc} = \frac{E_{Th} \angle \delta}{Z_{Th} \angle \alpha} = I_{k,sc} \angle \delta - \alpha \tag{5.63}$$

From Figure 5.11, the voltage equation and load equation are

$$V_k = E_{Th} - Z_{Th} \cdot I_k \quad \text{and} \quad V_k = Z_L \cdot I_k \tag{5.64}$$

From the equality of these equations, we get

$$E_{Th} - Z_{Th} \cdot I_k = Z_L \cdot I_k$$

or
$$E_{Th} - (Z_{Th} + Z_L) I_k = 0 \tag{5.65}$$

For maximum loading ($Z_L = Z_{Th}$). Therefore from Eq. (5.65),

$$E_{Th} - 2 Z_{Th} \cdot I_k^{cr} = 0$$

or
$$I_k^{cr} = \frac{E_{Th}}{2 Z_{Th}} = \frac{I_{k,sc}}{2} \quad \text{[from Eq. (5.63)]} \tag{5.66}$$

Hence, for maximum loading I_k^{cr} is always on the dotted line perpendicular to $\dfrac{\overline{I_{k,sc}}}{2}$ (Figure 5.12).

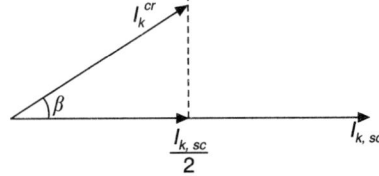

Figure 5.12 Current phasor.

Let us define: $-\beta = (\delta - \alpha) + \phi$ [also see Figure 5.14]
From Figure 5.12, critical current magnitude of bus k can be written as

$$I_k^{cr} = \frac{I_{k,sc}}{2\cos\beta} \quad (5.67)$$

since for maximum power transfer, $Z_L = Z_{Th}$, hence from Eq. (5.64)

$$V_k^{cr} = Z_L \cdot I_k = Z_{Th} \cdot I_k \quad (5.68)$$

Phasor diagram of Eq. (5.64) is given in Figure 5.13. It can be easily shown that $\bar{V}_k^{cr}$ is always on the dashed line perpendicular to $\dfrac{\bar{E}_{Th}}{2}$ with the magnitude of

$$\bar{V}_k^{cr} = \frac{\bar{E}_{Th}}{2\cos\delta} \quad (5.68a)$$

Using the above parameters and Eqs. (5.64), (5.57) and (5.68) we can construct the maximum power transfer phasor diagram given in Figure 5.14.

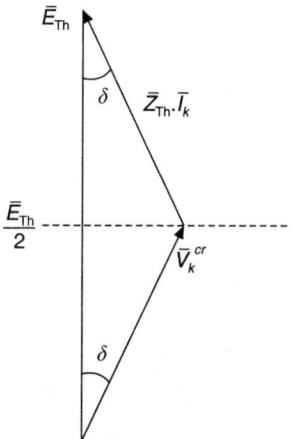

Figure 5.13 Voltage phasor.

To get the critical values, first it is required to fix some equalities, $v = \gamma - \beta$ and $v = \gamma - \delta$ which yields $\beta = \delta$. $\quad (5.69)$

Equation (5.69) shows that at maximum power transfer points, $\beta = \delta$. The resultant critical values of major parameters are

$$V_k^{cr} = \frac{E_{Th}}{2\cos\delta} \quad (5.70)$$

and

$$I_k^{cr} = \frac{I_{k,sc}}{2\cos\delta} \quad (5.71)$$

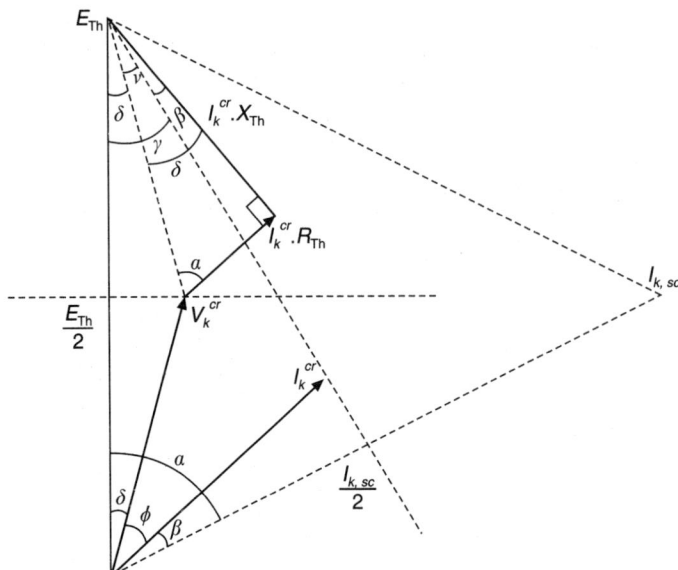

Figure 5.14 Maximum power transfer phasor diagram.

so, the critical power becomes

$$S_k^{cr} = \frac{S_{k,sc}}{4\cos^2 \delta} \qquad (5.72)$$

where $S_{k,sc}$ = short-circuit apparent power for bus k.

Here, we can introduce a new security margin as follows

$$\text{VSM} = \frac{I_k^{cr} - I_k}{I_k} \qquad (5.73)$$

By the use of this phasor diagram, for maximum loading condition, the angle β is to be equal to δ. All critical values evaluated are given as function of δ. Critical values obtained are not fixed and depend on the system Thevenin parameters and loading conditions. Therefore, critical values and VSM should be updated with online measurements and estimation.

5.11 A TOOL OF ASSESSING VOLTAGE STABILITY LIMIT

This section describes the basis (or fundamental) of using the *V–I* characteristic as a tool of assessing the voltage stability limit. For this, let us consider a general power system with a local load bus or voltage uncontrolled bus as shown in Figure 5.15(a). The rest of the system can be represented by its Thevenin equivalent (shown in Figure 5.15(b)). The load of the local bus is considered as $\bar{S} = |S| \angle \theta$ and is represented by a shunt impedance $\bar{Z} = |Z| \angle \theta$ in the equivalent system.

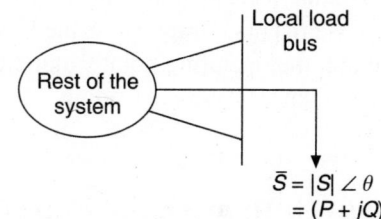

Figure 5.15(a) Multibus power system.

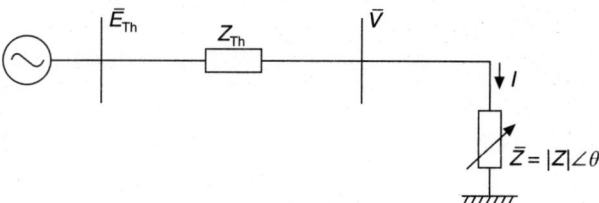

Figure 5.15(b) Thevenin equivalent of the multibus power system.

The load voltage, the current and power can be varied by changing the load impedance. A typical variation of load voltage magnitude against the load current (obtained by varying Z at a constant angle θ) is called the V–I characteristic of the bus. The complex load of the bus

$$\bar{S} = |S| \angle \theta = VI^* \qquad (5.74)$$

The magnitude of apparent power S is the product of load voltage and current magnitudes, i.e.

$$S = VI \qquad (5.75)$$

Let us consider an operating point x on the V–I characteristic (Figure 5.16). Thus the apparent power S_x is given by

$$S_x = V_x I_x \qquad (5.76)$$

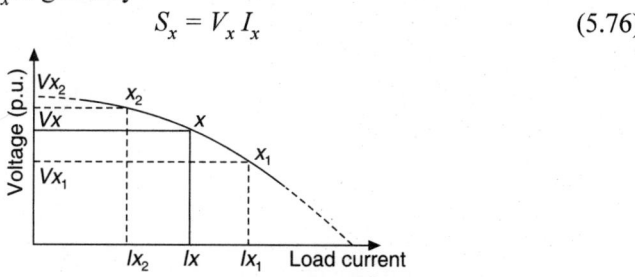

Figure 5.16 V – I characteristic.

Here, V_x and I_x are the load voltage magnitude and current magnitude respectively, at point x. It is to remember that, in a real power system, the V – I characteristic of any load bus can directly be generated from local measurements of voltage and current magnitudes. Thus, the knowledge of Thevenin parameters is not required at all to establish the V–I characteristic. One can easily recognize the maximum area along various paths or curves of Figure 5.16. One of the objectives of this analysis is to find the maximum load apparent power (or simply the critical load) at which the voltage collapse occurs. Graphically, the value of the critical load can be determined by sliding the operating point x along the V–I characteristic until the area becomes maximum. The critical loading value can be obtained from the maximum area.

5.12 EFFECT OF SYSTEM REACTANCE AND LOAD POWER FACTOR ON VOLTAGE STABILITY

The complex power flow equation of an uncompensated basic transmission system model is given by

$$S = VI^*$$

where

$$I = \frac{E - V}{X}$$

may be presented as

$$S = \frac{VE^*}{X^*} - \frac{VV^*}{X^*} \tag{5.77}$$

Taking V as the reference vector, simplification yields

$$S = \frac{EV}{X}\sin\delta + j\left(\frac{EV}{X}\cos\delta - \frac{V^2}{X}\right) \tag{5.78}$$

giving the well-known real and reactive power expressions of a lossless power transmission system as

$$P = \frac{EV}{X}\sin\delta \tag{5.79}$$

and

$$Q = \frac{EV}{X}\cos\delta - \frac{V^2}{X} \tag{5.80}$$

Elimination of δ results in the steady state voltage equation of the system represented by

$$V^4 + V^2(2Qx - E^2) + x^2(P^2 + Q^2) = 0 \tag{5.81}$$

Equation (5.81) is a quadratic equation in V^2, the solution being

$$V^2 = \left[\frac{-2Qx + E^2}{2} \pm \frac{1}{2}\sqrt{(2Qx - E^2)^2 - 4x^2(P^2 + Q^2)}\right]$$

Since the imaginary value of V carries no physical significance, the positive real root given by Eq. (5.82) has only been considered; thus

$$V = \left[\frac{-2Qx + E^2}{2} \pm \frac{1}{2}\sqrt{(2Qx - E^2)^2 - 4x^2(P^2 + Q^2)}\right]^{1/2} \tag{5.82}$$

For convenience, assuming unity power factor at the receiving-end bus, Eq. (5.82) becomes

$$V = \left[\frac{1}{2}E^2 \pm \frac{1}{2}\sqrt{E^4 - 4x^2P^2}\right]^{1/2} \quad (5.83)$$

Equation (5.83) signifies that the receiving-end voltage in an uncompensated transmission system is a function of the sending-end voltage (E), system reactance (x) and the receiving-end real and reactive powers (P, Q). Considering the simplest transmission model with voltage control at the sending end only, i.e. for $E = 1.00$ p.u. (constant), Eq. (5.83) reduces to

$$V = \left[\frac{1}{2} \pm \frac{1}{2}\sqrt{1 - 4x^2P^2}\right]^{1/2} \quad (5.84)$$

Both the real roots of V are equal when the expression under the radical sign is 0, i.e. when

$$x = \frac{1}{2P} = x_{cri} \quad \text{(say)} \quad (5.85)$$

Thus, from Eq. (5.83)

$$V = \left[\frac{1}{2} \pm \frac{1}{2}\sqrt{1 - \left(\frac{x}{x_{cri}}\right)^2}\right]^{1/2} \quad (5.86)$$

With $x < x_{cri}$, the expression under the radical sign is always real and for $x > x_{cri}$, the roots are imaginary without carrying any physical meaning.

It has been proposed to define the value of V with $x = x_{cri}$ in Eq. (5.84) or (5.85) as critical receiving-end voltage (V_{cri}) and is given by $V_{cri} = 0.707$ p.u. for an uncompensated lossless transmission system operating at unity power factor (UPF).

In EHV transmission systems, the lines being mostly reactive power constrained, the reactive power demand at the receiving end is expressed as

$$Q = P \tan \phi \quad (5.87)$$

where ϕ is the power factor at the receiving-end side. Substituting the value of Q from Eq. (5.87) in Eq. (5.80) and utilising Eq. (5.79), simplification yields

$$V = E(\cos \delta - \sin \delta \tan \phi) \quad (5.88)$$

However, the receiving-end voltage at voltage stability limit is given by

$$V_{cri} = \frac{E}{2 \cos \delta_{lim}}$$

when,

$$\delta_{cri} = \frac{\pi}{4} - \frac{\phi}{2} \quad (5.89)$$

At unity power factor, from Eq. (5.89), $\delta_{cri} = 45°$ when the magnitude of V_{cri} becomes 0.707 p.u. using Eq. (5.89) at voltage stability limit.

Hence, it can be inferred that the critical value of the receiving-end voltage V_{cri}, at the state when the system reactance is equal to the critical system reactance ($x = x_{cri}$), represents the voltage stability limit of a lossless basic power transmission system. Mathematically, voltage stability limit is obtained when the two real roots of the system voltage equation converge to a particular point and the Jacobian of the load flow equation becomes singular. Physically, the voltage stability limit can be defined as the limiting stage in a power system beyond which no amount of reactive power injection will elevate the system voltage to its nominal state. The system voltage can only be adjusted by reactive power injection till the system voltage stability is maintained. The critical receiving-end voltage as well as the bus voltage at the receiving end at any time are also governed by the load power factor as is evident from the above analysis and as shown below.

Expression for critical system reactance at voltage stability limit at any power factor

The expression for the critical system reactance at voltage stability limit for any power factor can be obtained by equating the expression with the radical sign of Eq. (5.82) to 0. Thus

$$(2Qx - E^2) = 4x^2(P^2 + Q^2)$$

which on simplification yields

$$4x^2 P^2 + 4x QE^2 - E^4 = 0 \quad (5.89a)$$

when

$$x = \frac{-4QE^2 \pm \sqrt{16Q^2 E^4 + 16E^4 P^2}}{8P^2} \quad (5.89b)$$

Using $Q = P \tan \phi$, simplification of Eq. 5.89(a) yields the value of the critical system reactance stability limit given by

$$x_{cri} = \frac{E^2}{2P}(-\tan \phi + \sec \phi) \quad (5.89c)$$

The following analysis describes another form of representation to show the effect of system susceptance on critical power angle at voltage stability limit.

In a lossless system, the reactive power flow expression can also be written as

$$V^2 B - VEB \cos \delta + Q = 0 \quad (5.90)$$

where B represents the system susceptance. Hence,

$$V = \left[\frac{E}{2} \cos \delta \pm \frac{1}{2} \sqrt{E^2 \cos^2 \delta - 4P \tan \phi \cdot \frac{1}{B}} \right] \quad (5.91)$$

Comparing Eq. (4.8) with Eq. (5.91), at voltage stability limit, simplification yields

$$\delta_{cri} = \cos^{-1}\left[\frac{E}{\left(2E^2 - 4P \cdot \frac{1}{B}\tan\phi\right)^{1/2}}\right] \quad (5.92)$$

Equation (5.92) represents the magnitude of δ_{cri}, the power angle at voltage stability limit, in terms of the sending-end voltage (E), real power (P), system susceptance (B) and receiving end (RPF/PF) [i.e. $\tan\phi$], and characterises the change in δ_{cri} for the change in B when the other parameters remain constant. In a particular case, when $E = 1.00$ (constant), at UPF, using Eq. (5.92) δ becomes 45°. This is identical to the result that has been obtained in the proceding analysis. This also helps to understand that Eqs. (6.80) and (6.83) are identical to each other, the only difference being the form of representation.

Figure 5.17 represents the characterisitics of the receiving-end voltage (V) of a basic power transmission system for a varying system reactance (x) or system short-circuit capacity in p.u. As the system short-circuit capacity is inversely proportional to the system reactance, it is evident from this figure that the receiving-end voltage falls with the increase in system reactance at any fixed real power demand of the system till the voltage stability limit is attained. For the lossless uncompensated line model, with $E = 1.00$ p.u., at unity power factor, the critical receiving-end voltage at voltage stability limit is determined as $V_{cri} = 0.707$ p.u. Both the real roots of Eq. (5.42) coincide at this critical point when $x = x_{cri}$, indicating the importance of critical system reactance on voltage limit. With

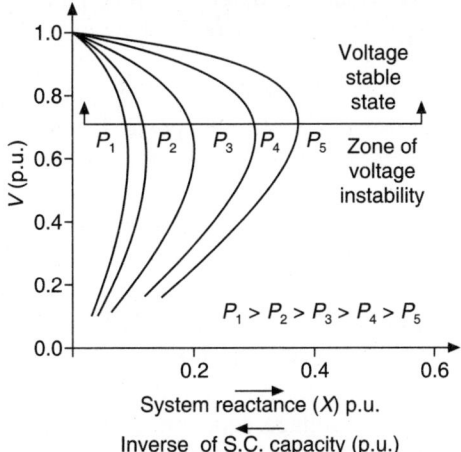

Figure 5.17 Receiving-end voltage profile and states of voltage stability against various system reactances at u.p.f. and at different load powers.

$x \neq x_{cri}$, two states of voltage operation are possible but the lower voltage operation needs a large current from the source at the rated and heavy loading condition and this is unacceptable. The higher value of the receiving-end voltage is termed the voltage stable state. The non-linearity of the voltage–reactance characteristic of Figure 5.17 near the 'knee' of the curves (i.e. at the voltage stability limit) is due to the sharp increase in transmission line current with drop in voltage for constant power operation leading to the rise of series reactive loss of the line given by the well-known relation $(dQ_L/dV) \propto (1/V^3)$ derived earlier. Here Q_L represents the series reactive loss. The effect is cumulative and may even lead to voltage collapse. Figure 5.18 indicates the characteristic of voltage vs. system short-circuit capacity for any fixed value of real power flow, considering 0.95, 0.9 power factors (lagging), u.p.f. and 0.95, 0.9 power factor (leading) load.

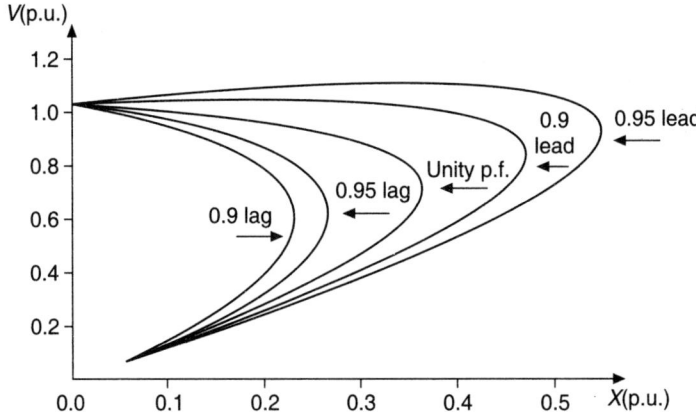

Figure 5.18 Receiving-end voltage profile for varying system reactances at different power factors for any specific amount of power flow.

Figure 5.19 represents the characteristics of δ_{cri}, the critical power angle at stability limit vs. x, the reactance at different power factors for any given amount of power flow. The decrease in the critical value of the power angle with an increase in system reactance offers an inherent limitation in the operation of the power system. This is because its voltage stability limit is attained at a much lower value of power angle following the increase in system reactance, that may be due to contingencies in addition to its high terminal reactance.

Hence in the basic power transmission model, an interesting criterion of voltage stability has just been presented. It has been found that the critical receiving-end voltage at voltage stability limit is given by the criterion of equality of the two real roots of the system voltage equation, and is governed by the critical system reactance at any specific amount of power flow. The voltage stability limit remains unaltered for any amount of power flow for any particular value of the receiving-end power factor. It has been revealed that due to low short-circuit capacity, which causes high system reactance, the critical receiving-end voltage

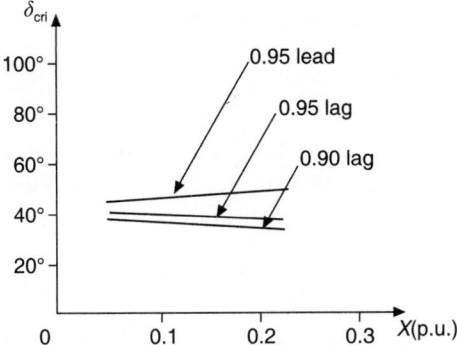

Figure 5.19 Profile of critical power angle at various power factors and different system reactances.

and power angle are low at stability limit in radial power systems. With operation at lagging power factor, the voltage stability limit and critical power angle further deteriorate.

This investigation also reveals that in an uncompensated system, the limiting value of the power angle is governed by the system reactance and load power factor. It can also be revealed that for a power system, the shunt capacitor compensation increases the critical power angle and the receiving-end voltage magnitude, at voltage stability limits. However, the voltage stability margin decreases with enhancement of shunt capacitive support. Proper selection of shunt capacititve support is thus desirable in order to have an acceptable voltage profile.

5.13 VOLTAGE STABILITY AND ITS RELATION WITH OFF-NOMINAL TAP RATIO OF LOAD SIDE TRANSFORMER AND SOURCE-TO-LOAD REACTANCES

Since on Load Tap Changing (OLTC) transformers are frequently used for voltage control at the load-end bus in power system, it should be kept in mind that OLTC transformers only adjust the reactive power flow between the buses. They do not produce any reactive power to assist the system when the system faces contingencies.

In this section analytical methods have been presented to determine the governing effect of the turns ratio on the On Load Tap Changer (OLTC) at the load bus of an EHV power transmission link and the ratio of source and load bus reactances on the receiving-end voltage, at stability limit under any specified operation criterion. The expressions developed in the text have been simulated on a typical EHV power transmission system to justify their validity.

In the simplest power system, the expression of the receiving-end voltage during maximum power transfer in a lossless frame is given by

$$V_{max} = \frac{E}{\sqrt{2(1+\sin\phi)}} \tag{5.93}$$

Under this condition, the amount of real power, that can be transferred, may be obtained as

$$P_{R_{max}} = \frac{V_{max}^2}{B} \cos \phi \qquad (5.94)$$

However, taking into account the load side transformer (Figure 5.20), the system voltage and current expressions are given by

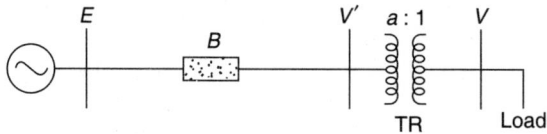

Figure 5.20 Schematic of a power system with receiving-end transformer with off nominal tap ratio.

$$\frac{V'}{V} = \frac{a}{1} = \frac{i}{I} \qquad (5.95)$$

when the modified power flow expressions in the lossless frame are given by

$$P_R = \frac{E(aV)}{B} \sin \delta \qquad (5.96)$$

and

$$Q_R = \frac{E(aV)}{B} \cos \delta - \frac{(aV)^2}{B} \qquad (5.97)$$

Hence, at voltage stability limit, equating the Jacobian to zero, the critical value of the receiving-end voltage is given by

$$V_{cri} = \frac{E}{(2 \cos \delta_{lim}) a} \qquad (5.98)$$

while the receiving-end voltage, under the normal operating condition, is given by

$$V = \frac{E}{a} (\cos \delta - \sin \delta \tan \phi) \qquad (5.99)$$

Considering the effect of the load bus transformer, the expression for the maximum power angle (at voltage stability limit and at any lagging power factor), is given by

$$\delta_{lim} = \frac{\pi}{4} - \frac{\phi}{2} \qquad (5.100)$$

Assuming $a = 1$, i.e. when the transformer's turns ratio becomes unity, Eq. (5.99) becomes $V = E (\cos \delta - \sin \delta \tan \phi)$.

Introducing a new term 'r', the ratio of source-to-load reactance of the lossless radial system, it can be expressed as

$$r = \frac{x_s}{x_{load}} = \frac{I \cdot x_s}{\frac{i}{a} \cdot x_{load}} = \frac{E - aV}{\frac{V}{a}}$$

where x_s is the summation of all reactances from the source to load bus and x_{load} is reactance of the load connected r the load bus. Thus,

$$r = \frac{a(E - aV)}{V}$$

i.e.
$$V = \frac{aE}{a^2 + r} \tag{5.101}$$

Hence, at voltage stability limit, for an optimum value of r, i.e. when $r_{opt} = (x_s/x_{load})_{opt}$ Eq. (5.101) can also be expressed as

$$V_{cri} = \frac{aE}{a^2 + r_{opt}} \tag{5.102}$$

Comparing Eq. (5.102) with Eq. (5.98), at the limiting stage of voltage stability, the optimum ratio of source-to-load reactance has been obtained as

$$r_{opt} = 2a^2 \cos\delta - a^2 \tag{5.103}$$

Equations (5.102) and (5.103) highlight the utility of low values of 'r'. This will result in a higher magnitude of the receiving-end voltage and greater amount of power transfer at voltage stability limit. This can only be achieved for systems with low system reactance (or high nodal short-circuit strength) at the source side.

Thus to have a voltage stable state, at the voltage uncontrolled load end of any reactive power constrained line, the operating state of the system can be analytically expressed as

$$V_{cri} \geq \frac{E}{(2\cos\delta_{lim})a} \tag{5.104}$$

where $\delta = (\pi/4 - \phi/4)$ and V_{cri} is also given by $V_{cri} = [aE/(a^2 + r_{opt})]$

with
$$r \leq 2a^2 \cos\delta_{lim} - a^2 \tag{5.105}$$

Denoting $(2a\cos\delta_{lim} - a^2) = L$, where L stands for the indicator of voltage stability, the voltage stability criterion can be represented as under:

$L > r$ for a voltage stable state,
$L = r$ at voltage stability limit,
$L < r$ during the stage of voltage instability.

From physical understanding, it is evident that, for voltage stable condition,

$$(dV/da) < 0 \tag{5.106}$$

Comparing Eqs. (5.101) and (5.106) at voltage stable state, $r_{opt} < a^2$, i.e. for voltage stability, $(x_s/x_{load}) < a^2$.

Assuming $a = 1$, to ensure voltage stability,

$$(x_s / x_{load}) < 1$$

This indicates, for normal operation $|x_{load}| > |x_s|$, i.e. the source impedance must be less than the load impedance. However, it may happen that, due to contingency, $|x_{load}| < |x_s|$. This causes voltage instability. The maximum power transfer is then given by $S_{max} = \dfrac{1}{4} \dfrac{E^2}{\cos^2 \delta_{lim} \cdot Z_{load}}$.

An increase in the value of the 'off-nominal tap ratio' does not improve the critical load-end voltage at any power factor, when the reactive power demand at the load bus is held constant. The optimum ratio of source-to-load impedance decreases with the improvement in power factor, and hence improves the operating condition at the voltage stability limit.

Thus it has been revealed that the critical receiving-end voltage of a longitudinal power link at the stability limit is governed by the 'off-nominal tap ratio and the load-side power factor. In case the load power factor deteriorates, the critical receiving-end voltage starts falling in magnitude and any attempt to boost up the off-nominal tap ratio deteriorates the load-end critical voltage further. Shunt compensation is not able to halt this characteristic since it can only improve the critical voltage magnitudes. Then, once the system is at the verge of voltage instability, in order to raise the load voltage, tap changer operation further complicates the situation without assisting it even if the load bus voltage is provided with shunt compensation. Improvement of load power factor is crucial to keep the load-end voltage at the stability limit. Obviously, improvement in load power as well as shunt compensation improves the maximum power transfer at the voltage stability limit.

Deterioration of power factor reduces the critical power angle and increases the optimum source-to-load impedance ratio. Thus, assuming the source impedance to be constant, the load side impedance must be reduced in order to ensure voltage stability once the load-side power factor drops. In weak power systems, where the source and the load bus impedances are inherently higher, voltage stability is severely threatened if the power factor of the load bus deteriorates. Operation of the on-load tap changer does not improve the situation as the improvements have been seen to be marginal.

Any decrement of source reactance also pushes the system closer to the point of instability. The limitation in capability of the longitudinal power system to operate at higher power factors restrains the maximum power transfer as well as the critical receiving-end voltage. Operation of an OLTC transformer at the voltage stability limit further pushes the system towards voltage instability.

EXERCISES

1. What is P–V curve? Why do we need a three-dimensional curve for assessing voltage stability?
2. What are the implications of representing the boundary of voltage stability in PQ plane?
3. Explain the significance of the Q–V curve in voltage stability.
4. Explain the concepts of dE/dV and dQ/dV criteria of voltage stability. What are their distinct features?
5. How would you take into account the magnitude of the load voltage as an indicator of stability?
6. Develop a direct indicator of voltage stability in a two-bus model. What is its effect on voltage stability margin?
7. Develop a voltage stability indicator based on governing effect of load increase.
8. Write short notes on:
   (i) L-index
   (ii) Singular value decomposition in forms of voltage stability indicators.
9. Derive the necessary equation in order to express the indicators of voltage stability as *FVSI* and *LQF*.
10. How do you use maximum power transfer theorem in voltage stability studies?
11. How do you assess voltage stability using V–I characteristics of a system?
12. Explain the role of system reactance and load power factor on voltage stability.
13. How does the operation of a *OLTC* affect the voltage stability at load bus? What is the effect of source reactance?

Chapter **6**

Assessment of Voltage Stability and Security

6.1 INTRODUCTION

The power system needs to be operationally secure, i.e. with minimal probability of blackout and equipment damage. An important component of power system security is the system's ability to withstand the effects of contingencies. A contingency is basically an outage of a generator, transformer and or line, and its effect is monitored with specified security limits. The power system operation is said to be normal when the power flows and bus voltages are within acceptable limits despite changes in load or available generation. From this perspective, security is the probability of a power system's operating point remaining in a viable state of operation.

Security assessment is a combination of system monitoring and contingency analysis. It is extremely uneconomical to build a power system with so much redundancy that failures will never cause an interruption of load on a system. Security assessment is analysis performed to determine whether, and to what extent, a power system is reasonably safe from serious interference to its operation. It involves the estimation of the relative robustness of the system in its present state or in the near future state. Direct security assessment indicates how the power system state will move from the normal operation state to the emergency state. Indirect security assessment can be formulated by defining a set of system security variables that should be maintained with predefined limits.

6.2 POWER SYSTEM SECURITY ANALYSIS

6.2.1 Analysis and Control of Operating States

The states of power system are classified into five states: normal, alert, emergency, extreme emergency and restorative. Figure 6.1 describes these states and the ways in which transition can occur from one state to another.

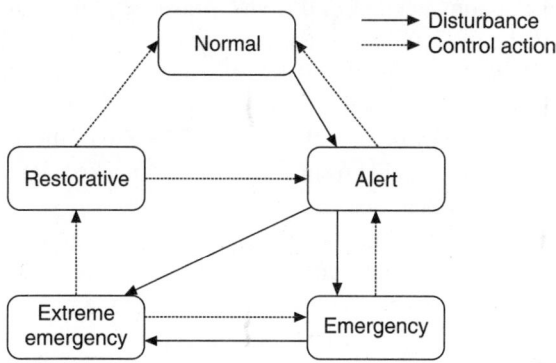

Figure 6.1 Power system operating states.

The operation of a power system in usually in a normal state. Voltages and frequency of the system are within the normal range and no equipment is overloaded in this state. The system can also maintain stability during disturbances considered in the power system planning. The security of the power system is described by thermal, voltage and stability limits. The system can also withstand any single contingency without violating any of the limits.

The alert state is similar to the normal state except that the above conditions cannot be met in the case of disturbance. The system transits into the alert state if the security level falls below a certain limit or if the possibility of a disturbance increases, e.g. because of weather conditions. All system variables are within the normal range and all constraints are satisfied in the alert operation point. However, the system has been weakened to a level where a contingency may cause a breaking of security limits. This will lead to a transition into an emergency state or in the case of severe disturbance into an extreme emergency state. Control actions, such as generation rescheduling, increased reserve, voltage control, etc. can be used to restore the system back to normal state. If this does not succeed, the system stays in the alert state.

The system transits into the emergency state if a disturbance occurs when the system is in the alert state. Many system variables are out of normal range or equipment loading exceeds short-term ratings in this state. The system is still complete. Emergency control actions, more powerful than the control actions related to alert state, can restore the system to alert state. The emergency control actions include fault clearing, excitation control, fast valving, generation tripping, generation run-back, HVDC modulation, load curtailment, blocking of on-load tap changer of distribution system transformers and rescheduling of line flows at critical lines.

The extreme emergency state is a result of the occurrence of an extreme disturbance or action of incorrect or ineffective emergency control actions. The system is in a state where cascading outages and shutdown of a major part of power system might happen. The system is in an unstable or close to an unstable state. The control actions needed in this state must be really powerful. Usually load shedding of the most unimportant loads and separation of the system into small independent parts are required in order to transit the system into a restorative state. This is only a possibility, because many production units are shut down due to too high or low frequency or any other reason. The aim of the actions is to save as much of the system as possible from a widespread blackout. If these actions do not succeed, the result is total blackout of the system.

The restorative state is a transition state between the extreme emergency and normal or alert states. It is important to restore the power system as fast and securely as possible in order to limit the social and economic consequences for the population and economy. The restoration time depends on the size of interrupted area, the type of production in the system, the amount of blackstart capability in the system and the possibility to receive assistance from interconnecting systems. The restoration process includes reconnection of all generators, transmission lines, loads, etc. Two major strategies for power system restoration following a blackout are the build-up and the build-down strategies. The build-up strategy is the most commonly used strategy. The system is divided into subsystems where each subsystem should have the capability to blackstart and control voltage and frequency. After the synchronisation of production units, loads are gradually connected. The connections to other subsystems are then synchronised.

Power system controls assist the operator in returning the system to a normal state after a disturbance. If the disturbance is small, this may be enough. Operator actions are required for a return to the normal state when the disturbance is large. The control system is highly distributed, which increases the reliability of control. Power system controls are organised as a hierarchical structure, such as

- direct control of individual components (excitation systems, prime movers, boilers, transformer tap changers, HVDC converters, Static VAR Compensators, automatically switched shunt compensation)
- an overall power or transmission plant control that co-ordinates the closely linked components
- a system control, which co-ordinates plant controls
- a pool-level master control, which co-ordinates system controls, and imports and exports of power flows.

6.2.2 Planning and Operation

Power systems are planned and operated so that most probable and critical contingencies can be sustained without an interruption or a quality reduction. The power system should be able to continue its operation despite sudden outage of a production unit, transmission line, transformer, compensation device, etc. Outages of power system equipment are typically due to faults (short-circuits and earth faults), overloads, malfunctions (false settings or operation actions) or breakdown of equipment. The occurrence of disturbances cannot be predicted, thus the security of the power system needs to be guaranteed beforehand.

The planning of a power system to ensure secure operation at minimum cost is a very complex problem with a potential for enormous financial gains in the solution. The security analysis is required to guarantee the power system's secure operation in all conditions and at all operation points. The purpose of power system planning is to ensure power system adequacy. The analysis is based on forecast cases. The on-line security assessment is needed in the power system operation to guarantee security momentarily, because not all possible future scenarios can be checked beforehand. The purpose of power system operation is to guarantee that a single disturbance cannot cause cascading outages and finally a total blackout.

The security of the power system commonly defined is based on $(n - 1)$ criteria, i.e. the system should withstand any single contingency. A system consisting of n components should be able to operate with any combination of $n - 1$ components, thus for any single component outage. This criterion plays an essential role in preventing major disturbances following severe contingencies. The use of this criteria ensures that the system will, at worst, transit from the normal state to the alert state.

6.2.3 Security Assessment

The information to assess power security is got from the Supervisory Control and Data Acquisition (SCADA) system. It provides the data needed to monitor the system state. Telemetry systems measure and transmit the data. The data consists of voltages, currents, power flows, the status of circuit breakers and switches, generator outputs and transformer tap positions in every substation in a transmission network. The state estimation is used in such a system to combine the raw measurement data with the power system model to produce the best estimate (in a statistical sense) of the power system state. Computers gather the data, process them, and store them in a database where operators can display and analyse them. The computer checks the incoming data against the security limits and notifies the operator in the case of overload or out-of-limit voltage.

The transmission network is controlled and optimised via a power management system. The power management system gets information from the SCADA system. It includes tools for operation point analysis (load-flow and dynamic simulation programs), contingency analysis (contingency screening and ranking), security assessment and security enhancement (optimal power flow program, e.g. to minimise network losses).

Figure 6.2 describes the off-line security assessment procedure found through security assessment software. The same procedure can be used in the on-line

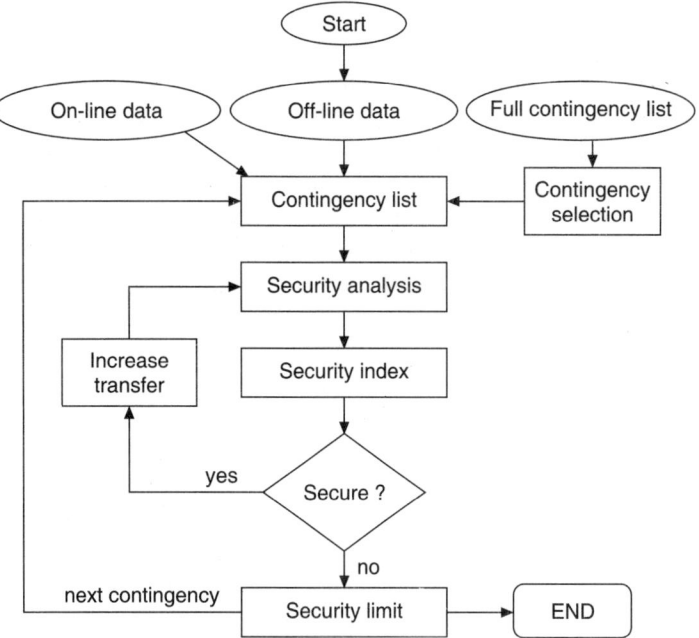

Figure 6.2 Schematic diagram of security assessment and determination of security limits.

security assessment, if the execution time is fast enough. The fundamental difference between off-line and on-line security assessment is the difference in input data. On-line security assessment analyses the current operation point given through the SCADA and the power management system, while the off-line security assessment ensures the security of a future operation point. It is based on the near future planning data, which must be predicted. The future operation points consider network topology, generation dispatch, unit commitment, load level, and power transactions. The uncertainty of these factors causes the need for numerous security assessment studies. As these possible variations increase, the number of operation points which must be studied becomes unmanageable, even in short-term planning cases. In on-line security assessment the variations of network topology, unit commitment and load level need not be considered. The number of possible variations to be studies becomes manageable and the stability limits obtained on-line are expected to be more accurate. However, the execution time of security assessment is much less in on-line than in off-line mode. The main features of the security assessment are contingency selection, security analysis using either load-flow or time domain simulation, and the determination of security limit. Contingency selection helps to identify the critical contingencies for the detailed security analysis. The analysis of all contingencies is too time-consuming and is not necessary in practice. Security analysis uses a short contingency list to check the security of the power system. The short contingency list includes those contingencies which are most probably critical. The traditional approach to planning for voltage security relied on ensuring that pre-contingency and post-contingency voltage levels were acceptable. This criterion is based on equipment tolerances and it ensures safe voltages. However, a system may have healthy pre-contingency and post-contingency voltage levels, but be dangerously close to voltage instability. That is why the security analysis should also provide an index to assure that sufficient voltage stability margin exists.

The security analysis is done using load-flow in the long-term studies and time-domain or fast time-domain simulation in the dynamic studies (quasi steady-state simulation, which is a compromise between the efficiency of load-flow methods and the accuracy of time-domain simulation). The algorithms must include detailed modelling capabilities and specific requirements, like variable step size integration, early termination, etc. to speed up the security analysis. The security of the system is usually indicated by a security index, e.g. voltage decline, reactive reserves, thermal overloads or voltage stability margin. The stress on the system can be increased if the system is secure. In this way the security limit, the last point where the system is still secure, is finally found. In practice, the security assessment is done every 5–10 minutes, because changes in operation points are normally small and slow, and long-term security analysis of practical size power system takes several minutes.

In the absence of on-line analysis capability, the off-line study results must be translated into operating limits and indices that can be monitored by the operators. The on-line security assessment is usually based on off-line computed security limits, e.g. transfer limits at tie-lines. The security limits are stored in a database and monitored on-line by matching the actual power system condition to the nearest matching condition in the database. These security limits are commonly presented to the operator by a security boundary.

The security boundary is a diagram giving the value of a quantity when a straight line is drawn between points, drawn between off-line computed security limits at the operation space of concerned parameters. In this way the security of power system can be monitored in a simple way. Secure operation is very probably guaranteed when the operation point is inside the security boundary. The number of interesting parameters is only two to three due to security boundary presentation problems.

Although the use of the security boundary method in the power system security assessment has provided major advantages especially in on-line applications, it has some drawbacks. The accuracy of security assessment in this way is not very good. The security limits tend to be conservative for normal conditions and may be inaccurate for unusual power system conditions. The maximum power flows and the minimum voltages cannot determine power system security explicitly in all possible network topology, loading and production situations, thus the security assessment is usually done for normal conditions and separate security limits are computed for maintenance periods. The accuracy of on-line security assessment is based on

- off-line computed security limits
- coverage of operation points in the operation space
- possibility to approximate power system security with monitored parameters.

6.3 COMPUTATION OF VOLTAGE STABILITY LIMITS

The computation of security limits in the case of voltage stability is described in this chapter, preceded by some definitions of transfer capacity and voltage stability margin. A system with large transfer capacity is generally more robust and flexible than a system with limited ability to accommodate power transfers. The transfer capacity is an indicator of power system security.

6.3.1 Concept of Transfer Capacity

The definition of transfer capacity is important in cases of congestion management and allocation of transfer capacity for market participants. Transmission system operators calculate the transfer capacity, the maximum power that can be transmitted between the systems in both directions. The definition of transfer capacity is distinguished in four terms (Figure 6.3):

- Total transfer capacity is set by physical limits of transmission systems. The security rules of transmission systems are taken into account in this capacity definition. The capacity limit is set by thermal, voltage and stability limits. The total transfer capacity represents the maximum feasible transfer between the systems, when considering security limits.
- Transmission reliability margin takes into account forecast uncertainties of power exchange. It is necessary to preserve a portion of the transfer capacity in order to ensure that the system remains secure under a wide range of varying conditions. The uncertainty is due to imperfect information from market actors (traders and generation companies) and unexpected events. Unexpected events, uncertain information and uncertainty about power system parameters can be considered as probabilistic events. Transmission system operators according to past experience or statistical methods do the calculation of transmission reliability margin.
- Net transfer capacity is the total transfer capacity minus the transmission reliability margin. It is the maximum transfer capacity that can actually be used and takes into account uncertainties in future power system conditions.
- Available transfer capacity is the net transfer capacity minus the notified transmission flow. The notified transmission flow is the transfer capacity already reserved by the contracts already accepted. In principle, a new contract of power exchange requires updating of net transfer capacity and decreases the uncertainty about future conditions. The available transfer capacity is the remaining transfer capacity after the prior allocations of contracted power exchanges. An example of transfer capacities is illustrated in Figure 6.3.

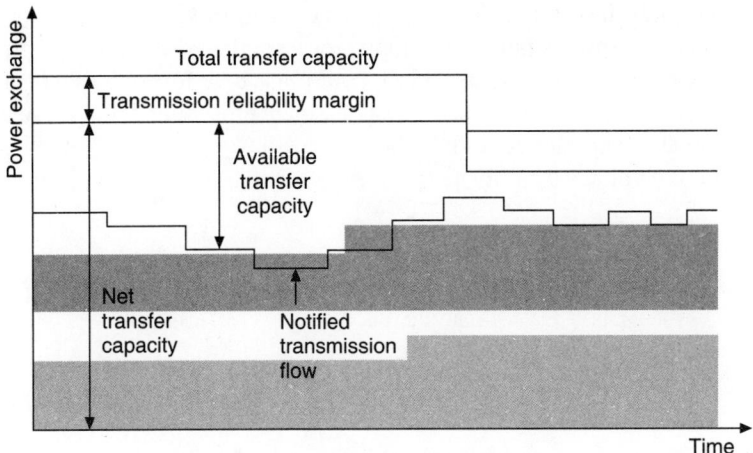

Figure 6.3 Example of transfer capacities.

The calculation of transfer capacity is complex in the case of highly interconnected and meshed systems. The calculation of total transfer capacity is based on modelling and simulation of the effect of power transfer between the systems. The simulation is started from the anticipated power system operation point. The total transfer capacity is calculated by shifting the generated power in order to cause additional power exchange between the systems. The power generation is increased at the export area and decreased at the import area. The load of both systems remains unchanged and the shift of generation is stopped when the security rules are not fulfilled in the systems. The total transfer capacity should be calculated in both directions. It is important to note that the calculation of the total transfer capacity is based on an assumption of a future generation dispatch scenario, load level and network topology. In order to handle these uncertainties, many total transfer capacities with different scenarios should be calculated.

6.3.2 Voltage Stability Margin

The voltage stability margin of a power system is a measure to estimate the available transfer capacity, net transfer capacity or total transfer capacity. The voltage stability margin is a difference or a ratio between the operation and voltage collapse points according to a key parameter (loading, line flow, etc.). The voltage collapse point must be assessed in order to guarantee secure operation at the normal operation point and after disturbances. The security of the power system is determined by the most critical post-disturbance margin. Figure 6.4 illustrates the voltage stability margins in the case of constant power load. The computation of voltage stability margin is time-consuming because many contingencies and post-disturbance margins should be studied. The pre-disturbance margin describes the loadability of a power system and is not interesting from the security point of view.

Theoretically the computation of the voltage stability margin should be done for all contingencies. This is a time-consuming process and is not necessary in practice. Thereby the most critical voltage stability margin is determined based on a short contingency list as in the general security assessment procedure. If the use of a short contingency list is not fast enough in the on-line voltage stability assessment, the computation of voltage stability margin also needs to be speeded up.

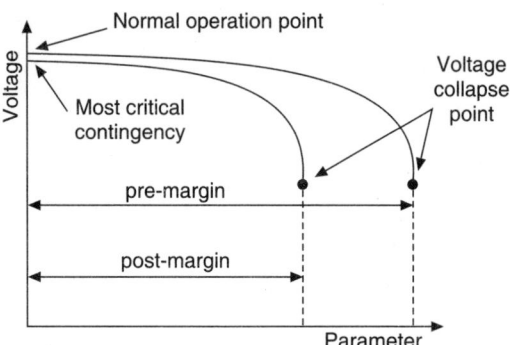

Figure 6.4 Voltage stability margins.

The voltage stability margin is a straightforward, widely accepted and easily understood index of voltage collapse. Other advantages of the voltage stability margin as a voltage collapse index are :

- The margin is not based on a particular power system model and can be used with static or dynamic models independent of the details of the power system dynamics.
- It is an accurate index that takes full account of the power system non-linearity and device limits as the loading is increased.
- Sensitivity analysis may be applied easily and quickly to study the effects of power system parameters and controls
- The margin accounts for the pattern of load increase (computation direction).

6.3.3 Computation of Voltage Collapse Point

Many good voltage stability indices and methods have been created for practical power systems. The most common methods to estimate the proximity of voltage collapse point are minimum singular value, point of collapse method, optimisation method and continuation load-flow. These are briefly described and discussed next. Other voltage stability indices are also proposed in the literature like sensitivity analysis, second order performance index , the energy function method, modal analysis and many others.

6.3.4 Method of Minimum Singular Value

The minimum singular value of load-flow Jacobian matrix has been proposed as an index for quantifying proximity to the voltage collapse point. It is an indicator available from normal load-flow calculations. The computation of minimum singular value can be done fast with a dedicated algorithm.

The singular value decomposition is applied to linearised load-flow equations to analyse power system voltage stability. The analysis studies the influence of a small change in the active and reactive power injections $[\Delta P \quad \Delta Q]^T$ on the change of angle and voltage $[\Delta \delta \quad \Delta V]^T$. The solution of linearised load-flow equations using the singular value decomposition is given in Eq. (6.1) where **J** is the load-flow Jacobian matrix.

$$\begin{bmatrix} \Delta \delta \\ \Delta V \end{bmatrix} = \mathbf{J}^{-1} \begin{bmatrix} \Delta P \\ \Delta Q \end{bmatrix} = (\mathbf{U \Sigma V}^T)^{-1} \begin{bmatrix} \Delta P \\ \Delta Q \end{bmatrix} = \mathbf{V \Sigma}^{-1} \mathbf{U}^T \begin{bmatrix} \Delta P \\ \Delta Q \end{bmatrix} \quad (6.1)$$

where U and V are $(m \times m)$ orthonormal matrices, Σ is a diagonal matrix (Σ = diag $\{r_i\}$) and r_i is singular value.

The inverse of the minimum singular value, $(\min\{\sigma_i\})^{-1}$, will indicate the greatest change in the state variables. Small changes in either matrix **J** or vector $[\Delta P \quad \Delta Q]^T$ may cause major changes in $[\Delta \delta \quad \Delta V]^T$, if min $\{\sigma_i\}$ is small enough. The minimum singular value is a measure of how close to singularity the load-flow Jacobian matrix is. If the minimum singular value of the load-flow Jacobian matrix is zero, then this matrix is singular, i.e. the inverse of the Jacobian matrix does not exist. The operation point does not have a load-flow solution in that case, because the sensitivity of the load-flow solution to small disturbances is infinite. The load-flow Jacobian matrix singularity point is also called the saddle node bifurcation point. Singular vectors can provide information about a power system's critical areas and components. The right singular vector corresponding to the smallest singular value indicates sensitive voltages and angles, i.e. critical areas. The left singular vector corresponding to the smallest singular value indicates sensitive direction for the changes in active and reactive power injections.

It should be remembered that a Jacobian matrix is a linearisation at the operation point and the voltage stability problem is non-linear in nature. If the operation point is far away from the voltage collapse point, then the minimum singular value does not describe the state of the system accurately. The minimum singular value of a load-flow Jacobian matrix is also sensitive to the limitations of generator reactive power, transformer tap changer and compensation device. The minimum singular value method may be applied to a reduced Jacobian matrix in order to improve the profile of the index.

6.3.5 Point of Collapse Method

The point of collapse method is a direct method to estimate the proximity of the voltage collapse point. It computes the voltage collapse point, the power demand and the corresponding state variables directly without computing the intermediate load-flow solutions. The method is based on the bifurcation theory and the singularity of the load-flow Jacobian matrix. In applying the bifurcation theory to power systems, the power demand is often used as the slowly changing parameter. A certain voltage collapse point is found in changes in parameters in a certain direction.

The voltage collapse point corresponds to the loss of equilibrium subject to smooth changes in parameters. The method is used to solve the singularity point of load-flow equations by formulating the set of equations which includes load-flow equations and equations describing conditions of singularity. Many types of bifurcations may occur, but the saddle node bifurcation is the most common in power systems. This occurs at the equilibrium when the corresponding load-flow Jacobian matrix **J** is singular.

The equations of saddle node bifurcation can be solved by, e.g. Newton–Raphson iteration techniques. Although the load-flow Jacobian matrix is singular at the bifurcation point, the Jacobian matrix is non-singular at the bifurcation point. The Jacobian matrix of equations is needed to calculate symbolically or numerically. The use of the Newton–Raphson method requires good initial values in order to converge to the bifurcation point. The system might have many bifurcation points, but good initial values for the eigenvector, guarantees convergence to the nearest voltage collapse point. A good guess for an initial eigenvector is an eigenvector corresponding to the smallest eigenvalue at the base case. The initial values for the state variables are the state values of load-flow solution at the base case. If the voltage collapse point is far from the base case, convergence problems may emerge. In that case new initial values are computed closer to the collapse point.

When applying the point of collapse method to voltage stability analysis, the information included in the eigenvectors can be used in the analysis of voltage stability. The right eigenvector defines the buses close to voltage collapse. The biggest element of the right eigenvector shows the most critical buses. The left eigenvector defines the bifurcation point normal vector. The selection of the opposite direction of the normal vector is the best way to improve the voltage stability at the collapse point. One of the problems of the point of collapse method is convergence to the collapse point from the distant initial point, when the reactive power limits of few generators are hit.

6.3.6 Optimisation Method

Another direct computation method of voltage collapse point is the optimisation method. In that case the voltage stability margin is maximised according to load-flow equations and power system constraints.

The solution of the optimisation problem satisfies the optimality conditions known as Khun–Tucker conditions.

The solution of non-linear equations can be solved using the Newton method. The load-flow Jacobian matrix is part of the Hessian matrix on the linear system. Although the load-flow Jacobian matrix is singular at the solution, the Hessian matrix is not. The vector requires good initial values in order to achieve quick and reliable convergence to the maximum point.

Inequality constraints, like generator reactive power limits, can be included by computing a sequence of optimisation problems. First, the unconstrained optimisation problem described above is solved. In that case all generators control their voltage as in the base case. The solution of the optimisation problem overestimates the real voltage stability margin. Then it is estimated, e.g. using the linear interpolation technique, which generators are switched to operate at the reactive power limit. A new optimisation problem starting from the previous solution is computed. The procedure is repeated until all constraints are satisfied. The final solution is the voltage stability margin.

6.3.7 Continuation Load-flow Method

The purpose of continuation load-flow is to find a continuum of load-flow solutions for a given load/generation change scenario, i.e. computation direction. It is capable of solving the whole PV-curve. The singularity of continuation load-flow equations is not a problem; therefore the voltage collapse point can be solved. The continuation load-flow finds the solution path of a set of load-flow equations that are reformulated to include a continuation parameter. This scalar equation represents the phase conditions that guarantee the non-singularity of the set of equations. The method is based on the prediction-correction technique. The prediction-correction technique applied to the PV-curve solution is illustrated in Figure 6.5. The intermediate results of the continuation process also provide valuable insight into the voltage stability of the system and the areas prone to voltage collapse.

The prediction step estimates the next PV-curve solution based on a known solution. Taking an appropriately sized step in a chosen direction to the solution path can make the prediction of the next PV-curve solution. However, the prediction is not necessary, especially at the flat part of the PV-curve. The simplest prediction is the secant the of last two PV-curve solutions. The computation of the secant is fast and simple. The tangent of the last PV-curve solution is more accurate than the secant, but also requires more computation. The advantage of the tangent direction is most valuable around the PV-curve nose. The step size should be chosen so that the predicted solution is within the radius of convergence of the corrector. The determination of the step size can be based on the slope of tangent or the difference between the previous predicted and exact solutions.

The inexact estimate is corrected using the slightly modified load-flow equations at the correction step which is based on, e.g. local parameterisation, where one state variable is constant during the computation or correction step. In Figure 6.5 the constant state variable is the continuation parameter at the first two corrections and the voltage at the last correction. There is no restriction for the continuation parameter, but usually it is the increment of total load.

The maximum loading point can be sensed easily using the tangent vector of the PV-curve solution. The tangent component corresponding to the continuation parameter is zero at the maximum loading point and becomes negative beyond the maximum point. This method indicates whether or not the maximum loading point has been passed. The exact location of the maximum loading point requires searching with a decreasing step size around the maximum point, which is why

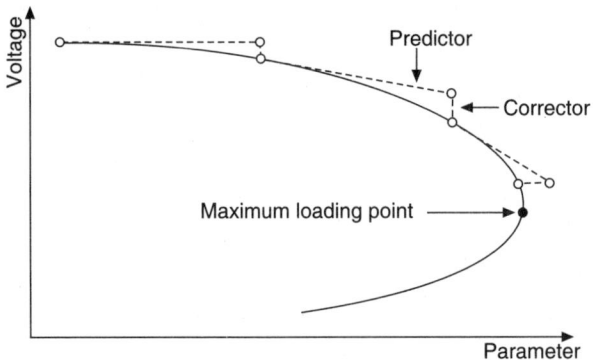

Figure 6.5 *PV*-curve solution using prediction-correction technique.

the point of collapse method and the optimisation method are more effective than the continuation load-flow to find the exact voltage collapse point.

The tangent vector can be used for a sensitivity analysis so that a stability index and identification of weak buses are obtained. A good method to decide which bus is nearest to its voltage stability limit is to find the bus with the largest ratio of differential change in voltage to differential change in active load for the whole system. This ratio can also be used as a voltage stability index.

6.3.8 Computation of Maximum Loading Point

The maximum loading point can be attained through a load-flow program. The maximum loading point can be calculated by starting at the current operating point, making small increments in loading and production and re-computing load-flows at each increment until the maximum point is reached. The load-flow diverges close to maximum loading point because there are numerical problems in the solution of load-flow equations.

Computation algorithm

The computation of the maximum loading point is a parametric analysis of the operating points. The computation algorithm is presented in Figure 6.6. The algorithm is based on the prediction-correction type computation, where loads and generation are increased towards the chosen computation direction. In this way, the n-dimensional boundary of maximum loading points can be determined in a certain direction. The solution of load-flow corrects the errors due to the discrepancy between the prediction and the actual non-linear functioning of the network. The convergence of load-flow is used as a criterion for the acceptance of intermediate results.

The increment of loads and generation is continued until the load-flow diverges. The algorithm is then continued with a decreased step length. The stopping criterion for the algorithm is the minimum step length. The accuracy of the proposed method depends on the value of the minimum step length. The flat part of the *PV*-curve can be computed fast. The first few steps may be large, but near the maximum loading point, it must be reduced significantly due to load-flow numerical problems. The computation time is mostly spent at diverged load-flows close to the maximum loading point. The computation time is minimised when the number of diverged load-flows is minimised. The proposed method has proved to reach almost the same solution as the continuation load-flow with constant power loads.

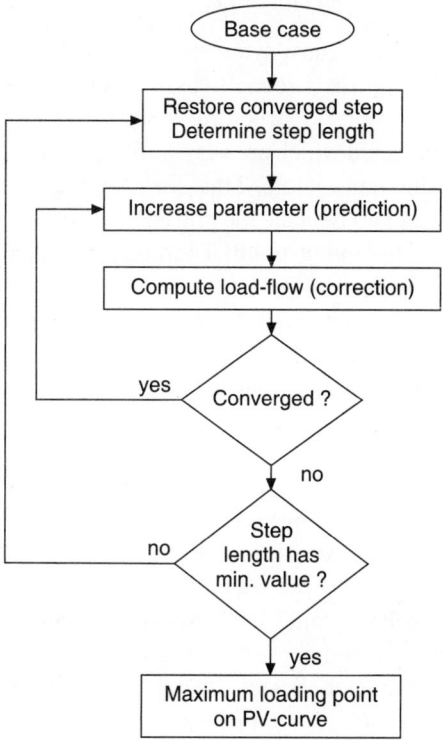

Figure 6.6 Algorithm of maximum loading point computation.

6.3.9 Eigenvalue Analysis

Eigenvalues are also often used to determine proximity to the voltage collapse point. Its application to voltage collapse analysis focuses on monitoring the smallest singular value up to the point when it becomes zero at the collapse point. It may be noted here that minimum eigenvalue is the relative measure of how close the system is to the voltage collapse point or singular point of load flow Jacobian matrix.

The Jacobian matrix used to compute bus voltages in Newton-Raphson load flow method is given by

$$\begin{bmatrix} \Delta P \\ \Delta Q \end{bmatrix} = \begin{bmatrix} J_1 & J_2 \\ J_3 & J_4 \end{bmatrix} \begin{bmatrix} \Delta \delta \\ \Delta |V| \end{bmatrix} \qquad (6.2)$$

With $\Delta P = 0$, we can write from Eq. (6.2),

$$[\Delta Q] = [J_R] \Delta |V| \qquad (6.3)$$

where, $\qquad [J_R] = [J_4] - [J_3][J_1]^{-1}[J_2] \qquad (6.4)$

$[J_R]$ is the reduced Jacobian matrix of the system.
From Eq. (6.3),

$$[\Delta |V|] = [J_R]^{-1} [\Delta Q] \qquad (6.5)$$

$[J_R]$ as defined in Eq. (6.5), is quasi-symmetric and therefore diagonalizable. This decomposition may be applied directly to assess voltage stability. In addition to that due to its quasi-symmetric structure, a set of only real eigenvalues and eigenvectors can be expected which are very similar in value to the corresponding singular values and singular vectors. Thus for $[J_R]$, the eigenvectors associated with eigenvalues closest to 0 indicate the presence of voltage collapse point in the vicinity of the present operating point. Therefore the maximum entries in the right eigenvector correspond to the critical buses (most sensitive buses) in the system and the maximum entries in the left eigenvector pinpoint the most sensitive direction for change of power.

The ith diagonal element of the matrix $[J_R]^{-1}$ also indicates the V–Q sensitivity load bus-i, when $\dfrac{d|V_i|}{dQ_i} = [J_R]^{-1}_{ii}$.

The V–Q sensitivity at load bus represents the slope of the V–Q curve at the given operating point. *A positive V–Q sensitivity is an indicator of stable operation, the smaller the sensitivity, the more stable is the bus.* As the system crawls in the vicinity of vulnerable voltage collapse, the sensitivity increases and at the voltage stability limit it becomes infinite.

Voltage stability characteristics of the load buses in a power system can be assessed by computing a set of eigenvalues and the corresponding eigenvectors of the reduced Jacobian matrix $[J_R]$ given in Eq. (6.2).

Let $\qquad [J_R] = [\xi][\lambda][\eta] \qquad (6.6)$

where $[\xi]$ = right eigenvector of $[J_R]$, $[\eta]$ = left eigenvector of $[J_R]$ and $[\lambda]$ = diagonal eigenvalue matrix of $[J_R]$.

Combining Eqs. (6.5) and (6.6),

$$\Delta[|V|] = [\xi][\lambda]^{-1}[\eta][\Delta Q] \qquad (6.7)$$

$\therefore \qquad [\Delta|V|] = \left[\sum_i \dfrac{\xi_i \eta_i}{\lambda_i} \Delta Q \right] \qquad (6.8)$

where, ξ_i is the ith column right eigenvector and η_i is the ith row left eigenvector of $[J_R]$. Each eigenvalue λ_i and the corresponding right and left eigenvectors ξ_i and η_i define the ith mode of Q–V response. The voltage stability analysis using the eigenvalue technique is also known as *Modal Analysis*.

From Eq. (6.8) it is clear that as λ_i tends to zero, $\Delta[|V|]$ tends to infinite. Therefore at voltage stability limit $\lambda_i = 0$.

V–Q sensitivity

From Eq. (6.8), V–Q sensitivity at load bus-k can be written as

$$\frac{\partial |V_k|}{\partial Q_k} = \sum_i \frac{\xi_{ki}\, \eta_{ik}}{\lambda_i} \qquad (6.9)$$

Thus using the eigenvalue analysis, the V–Q sensitivity can also be computed. The bus participation factor (P_{ki}) can also be computed from modal analysis where,

$$\therefore \quad [P_{ki} = \xi_{ki}\, \eta_{ik}] \qquad (6.10)$$

The bus participation factors obtained from modal analysis are also very effective to identify the zone vulnerable to voltage collapse. The bus participation factors also indicate the *degree of weakness of a bus in a particular mode*. The higher the bus participation, the weaker the load bus and it gradually leads towards voltage instability.

A simple flow chart to compute V–Q sensitivity and bus participation factors of a power supply system using modal analysis with two weakest modes (i.e. two minimum eigenvalues of reduced Jacobian matrix), is shown in Figure 6.7.

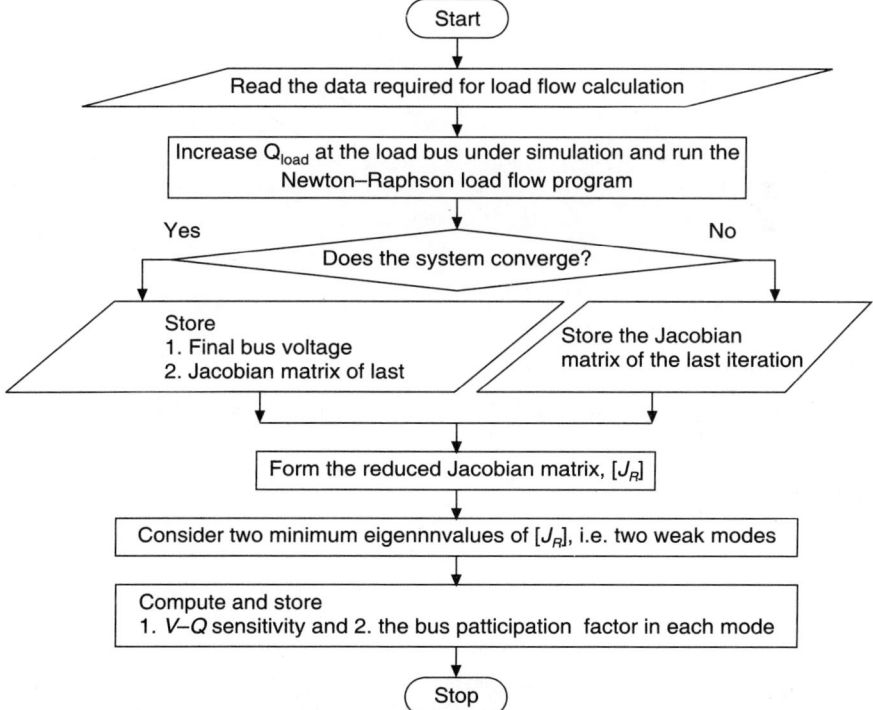

Figure 6.7 Flow chart to compute V–Q sensitivity and bus participation factors of a power supply system using modal analysis.

6.3.10 Bus Equivalencing Technique

The static load flow equations for a multi-bus power network are given by

or

$$\begin{bmatrix} \Delta\delta \\ \dfrac{\Delta|V|}{|V|} \end{bmatrix} = \begin{bmatrix} \dfrac{\partial P}{\partial \delta} & |V|\dfrac{\partial P}{\partial |V|} \\ \dfrac{\partial Q}{\partial \delta} & |V|\dfrac{\partial Q}{\partial |V|} \end{bmatrix}^{-1} \begin{bmatrix} \Delta P \\ \Delta Q \end{bmatrix} \quad (6.11a)$$

$$\begin{bmatrix} \Delta\delta \\ \dfrac{\Delta|V|}{|V|} \end{bmatrix} = [J]^{-1} \begin{bmatrix} \Delta P \\ \Delta Q \end{bmatrix} \quad (6.11b)$$

where $[J]$ is the Jacobian being represented in Eq. (6.11a). The voltages at the load buses being obtained from the above load flow equation, the power flow in each line is obtained from Eq. (6.12).

and
$$P_{ik} - jQ_{ik} = V_i^* (V_i - V_k) y_{ik} + V_i^* V_i (y'_{ik}/2)$$
$$P_{ki} - jQ_{ki} = V_k^* (V_k - V_i) y_{ki} + V_k^* V_k (y'_{ik}/2) \quad (6.12)$$

The power loss of the entire network, being the algebraic sum of all line flows in the system, is therefore given by

$$P_{\text{loss multibus}} = \sum_{i=1}^{N} (P_{ij} + P_{ji}) \quad (6.13a)$$

$$Q_{\text{loss multibus}} = \sum_{i=1}^{N} (Q_{ij} + Q_{ji}) \quad (6.13b)$$

where N = total number of buses in the system.

Next, consider a two-bus network (Figure 6.8) connected by a line impedance representing the equivalent model of the entire multi-bus network. In order to derive such a two-bus network we consider the sending end quantities having source power P_g and Q_g while the receiving end loads are P_{load} and Q_{load}. The power flow equation for such an equivalent network can be represented as

$$P_g = P_{\text{loss}} + P_{\text{load}} \quad (6.14a)$$
and
$$Q_g = Q_{\text{loss}} + Q_{\text{load}} \quad (6.14b)$$

Here we assume that the total transmission line loss is same both in multibus system and equivalent two-bus system. The real and reactive power losses (P_{loss} and Q_{loss}) for this equivalent system are then given by

Assessment of Voltage Stability and Security

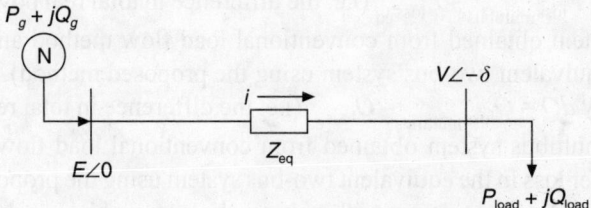

Figure 6.8 Equivalent two-bus system.

$$P_{loss} = r_{eq}\left(\frac{P_g^2 + Q_g^2}{E^2}\right) \tag{6.15a}$$

and

$$Q_{loss} = x_{eq}\left(\frac{P_g^2 + Q_g^2}{E^2}\right) \tag{6.15b}$$

where E being the sending-end voltage, r_{eq} and x_{eq} represent the equivalent resistance and reactance of the two-bus network.

The equivalent impedance of the two-bus network is then given by

$$z_{eq} = r_{eq} + jx_{eq} \tag{6.16}$$

where

$$r_{eq} = \frac{P_{loss}}{P_g^2 + Q_g^2} = \frac{P_g - P_{load}}{P_g^2 + Q_g^2} \tag{6.17a}$$

and

$$x_{eq} = \frac{Q_{loss}}{P_g^2 + Q_g^2} = \frac{Q_g - Q_{load}}{P_g^2 + Q_g^2} \tag{6.17b}$$

The sending-end voltage E being assumed to be at nominal value ($E = 1.0$ p.u.), P_{loss} and Q_{loss} in Eqs. (6.17a) and (6.17b) can be obtained from the expression of P_{loss} and Q_{loss} in Eqs. (6.14a) and (6.14b) respectively.

The receiving-end voltage, V, can easily be calculated as shown below:

$$V = E - z_{eq}\frac{P_g - jQ_g}{E} \tag{6.18}$$

where the transmission current is given by,

$$i = \frac{E - V}{z_{eq}} \tag{6.19}$$

The new values of system losses are

$$P_{loss_{eq}} = i^2 r_{eq} \tag{6.20a}$$

and

$$Q_{loss_{eq}} = i^2 x_{eq} \tag{6.20b}$$

In order to check the validity of the two-bus equivalent at any particular load level, the difference in total transmission line loss between the multi-bus system and the equivalent two-bus system is needed.

Let $dP = P_{loss_{multibus}} - P_{loss_{eq}}$ (i.e. the difference in total real power loss in the multibus system obtained from conventional load flow method and real power loss in the equivalent two-bus system using the proposed method).

Similarly $dQ = Q_{loss_{multibus}} - Q_{loss_{eq}}$ (i.e. the difference in total reactive power loss in the multibus system obtained from conventional load flow method and reactive power loss in the equivalent two-bus system using the proposed method).

If dP or $dQ < \varepsilon$, a tolerance, then it can be reasonably concluded that this method represents an equivalent two-bus model of the multibus system. The two-bus system thus developed above becomes the equivalent model of a multibus network at any particular network and load configuration.

Next, forming the Jacobian of the power flow equation and equating it to 0, (the Jacobian becomes singular when $\Delta[J] = 0$, signifying the singularity of the Jacobian being coincident with the voltage stability limit) the expression for critical receiving-end voltage (V_{cri}) becomes

$$|V_{cri}| = \frac{|E|}{2\cos\delta_{cri}} \tag{6.21}$$

Also, here
$$\delta_{cri} = \frac{\pi}{4} - \frac{\phi}{2} \tag{6.22}$$

Again taking V as the reference phasor, the fundamental power equation may be represented as

$$V^2 - VE + S^*_{load_{eq}} z_{eq} = 0 \tag{6.23}$$

Taking the first derivative of above equation with respect to $S_{load_{eq}}$ (complex load bus power),

$$2V\frac{dV}{dS_{load_{eq}}} - E\frac{dV}{dS_{load_{eq}}} + z_{eq} = 0$$

Simplification yields (at voltage stability limit),

$$\frac{dV}{dS_{load_{eq}}} = \frac{z_{eq}}{E - 2V} \tag{6.24}$$

$$\therefore \quad \left|\frac{dV}{dS_{load_{eq}}}\right| = \frac{|z_{eq}|}{|E| - 2|V|}$$

Taking into account the profile of voltage vs. $S_{load_{eq}}$ (Figure 6.9), voltage stability limit is attained obviously when $\frac{dV}{dS} = 0$; voltage stability is maintained till the slope of $\frac{dV}{dS}$ is negative (i.e. when $V < E/2$).

Therefore, with $E = 1$ p.u., voltage stability of the two-bus network is maintained when

$$\left|\frac{dV}{dS_{load_{eq}}}\right| = \frac{|z_{eq}/2|}{0.5 - V} \tag{6.25}$$

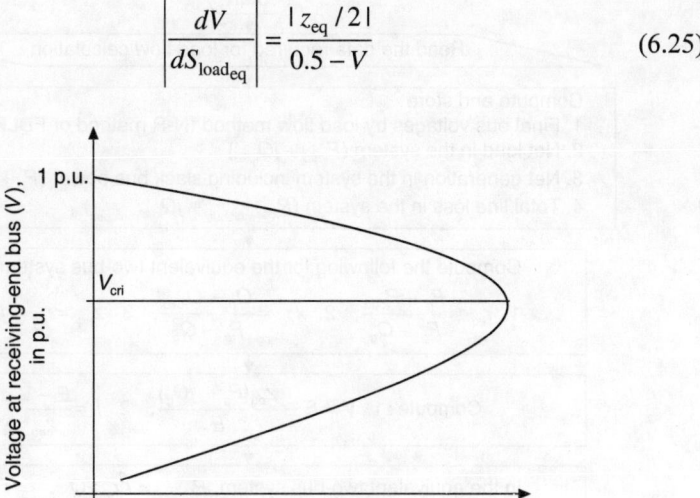

Figure 6.9 Profile of $|V|$ vs. $|S_{load_{eq}}|$ in a two-bus power system.

The flow chart of this method of voltage stability assessment is shown in Figure 6.10.

6.3.11 Method of Optimal Power Flow

The cost function ($F_{c_{total}}$) of an N-bus power system having NG number of fossil fuel units is given by

$$F_{c_{total}} = \sum_{i=1}^{NG} F_{c_i} = \sum_{i=1}^{NG} \alpha_i P_{g_i}^2 + \beta_i P_{g_i} + \gamma_i \text{ unit of cost/hr}$$

where

α_i, β_i and γ_i are the cost coefficients and

P_{g_i} = real power generation at the ith generator bus

NG = total number of generators

F_{c_i} = cost of generation at the ith generator

$F_{c_{total}}$ = total cost of generation in the system.

In this optimization problem, the network losses are neglected. The aim of optimal power flow (OPF) is to optimise the cost function subjected to the following constraints:

(i) *Active power balance in the network*

$$P_i(|V|, \delta) - P_{g_i} + P_{load_i} = 0 \text{ for } i = 1, 2, 3, \ldots, N$$

where P_i = active power injection at the ith bus and is a function of $|V|$ and δ. For load buses [i.e. for i = NG + 1), (NG + 2), ..., N], P_{g_i} = 0;

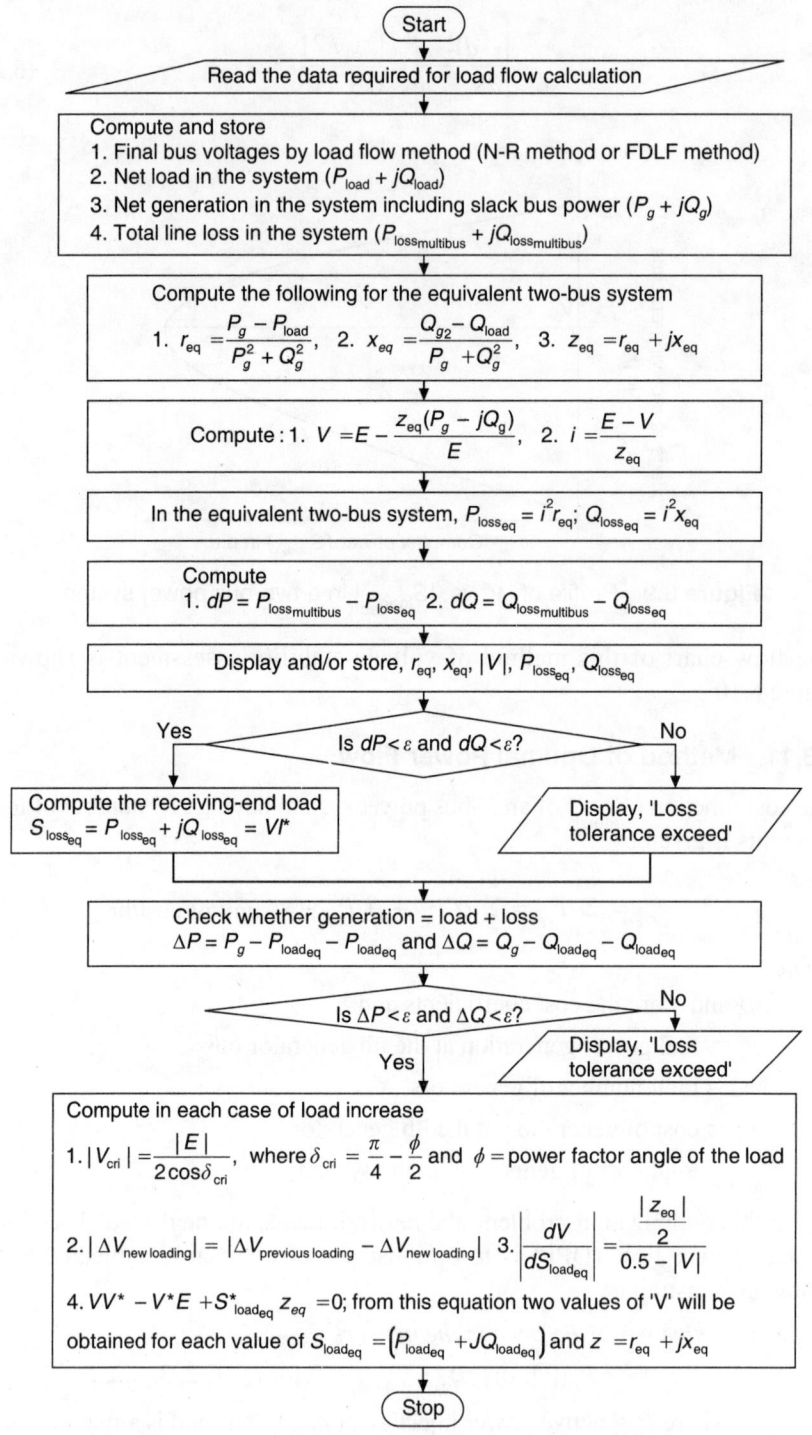

Figure 6.10 Flow chart of equivalencing a multibus system and computing voltage stable states.

(ii) *Reactive power balance in the network*

$$Q_i(|V|, \delta) - Q_{g_i} + Q_{\text{load}_i} = 0 \quad \text{for } i = (NG + 1), ..., N$$

where Q_i = reactive power injection at the ith load bus and is also a function of $|V|$ and δ; Q_{g_i} = reactive power generation at the ith bus;

(iii) *Security related constraints* (also called *soft constraints*); these constraints are as follows:

    (a) Limits on real power generations, i.e.

$$P_{g_{i\min}} \leq P_{g_i} \leq P_{g_{i\max}} \quad \text{for } i = 1, 2, 3, ..., NG$$

    (b) Limits on reactive power generations, i.e.

$$Q_{g_{i\min}} \leq Q_{g_i} \leq Q_{g_{i\max}} \quad \text{for } i = 1, 2, 3, ..., NG$$

    (c) limit on voltage magnitudes of load buses

$$|V_i|_{\min} < |V_i| < |V_i|_{\max} \quad \text{for } i = (NG + 1), ..., N \text{ and}$$

    (d) limit on voltage angles of all of the above buses excluding slack bus

$$\delta_{i\min} \leq \delta_i \leq \delta_{i\max} \quad \text{for } i = 2, 3, ..., N.$$

The real and reactive power injection at bus-i can be expressed as

$$P_i = \sum_{k=1}^{N} |V_i V_k| [G_{ik} \cos(\delta_i - \delta_k) + B_{ik} \sin(\delta_i - \delta_k)] \qquad (6.26a)$$

$$Q_i = \sum_{k=1}^{N} |V_i V_k| [G_{ik} \sin(\delta_i - \delta_k) - B_{ik} \cos(\delta_i - \delta_k)] \qquad (6.26b)$$

The constraint minimisation problem can be transformed into an unconstrained one by augmenting the load flow constraints into an objective function. The additional variables are known as *Lagrange multiplier functions (or incremental cost functions)* in power system optimisation. The Lagrangian function then becomes

$$L(P_g, |V|, \delta) = \sum_{i=1}^{NG} F_c(P_{g_i}) + \sum_{i=1}^{N} \lambda_{p_i} [P_i(|V|, \delta) - P_{g_i} + P_{\text{load}_i}]$$

$$+ \sum_{i=NG+1}^{N} \lambda_{q_i} [Q_i(|V|, \delta) - Q_{g_i} + Q_{\text{load}_i}] \qquad (6.27)$$

where λ_{p_i} and λ_{q_i} are Lagrangian multiplier for active power and reactive power balance at the ith bus respectively.

The optimisation problem can be solved, only if the following equations are optimally satisfied:

$$\frac{\partial L}{\partial P_{g_i}} = \frac{\partial F}{\partial P_{g_i}} - \lambda_{p_i} \quad \text{for } i = 1, 2, 3, ..., NG \qquad (6.28a)$$

$$\frac{\partial L}{\partial \delta_i} = \sum_{k=1}^{N} \left[\lambda_{p_k} \frac{\partial P_k}{\partial \delta_i}\right] + \sum_{k=1}^{N} \left[\lambda_{q_k} \frac{\partial Q_k}{\partial \delta_i}\right] \quad \text{for } i = 2, 3, ..., N \qquad (6.28b)$$

Also from Eq. (6.27), we can write

$$\frac{\partial L}{\partial \lambda_{p_i}} = P_i(|V|,\delta) - P_{g_i} + P_{\text{load}_i} \quad \text{for } i = 1, 2, 3, \ldots, N \qquad (6.28c)$$

and
$$\frac{\partial L}{\partial |V_i|} = \sum_{k=1}^{N}\left[\lambda_{p_k}\frac{\partial P_k}{\partial |V_i|}\right] + \sum_{k=NG+1}^{N}\left[\lambda_{q_k}\frac{\partial Q_k}{\partial |V_i|}\right] \quad \text{for } i = NG+1, \ldots, N$$
$$(6.28d)$$

Further to this,

$$\frac{\partial L}{\partial \lambda_{q_i}} = Q_i(|V|,\delta) - Q_{g_i} + Q_{\text{load}_i} \quad \text{for } i = (NG+1), \ldots, N \qquad (6.28e)$$

Any small variation in control variables about their initial values can be obtained by forming total differentials as shown below:

$$\sum_{k=1}^{NG}\frac{\partial^2 L}{\partial P_{g_i}\partial P_{g_k}}\Delta P_{g_k} + \sum_{k=2}^{N}\frac{\partial^2 L}{\partial P_{g_i}\partial \delta_k}\Delta \delta_k + \sum_{k=1}^{N}\frac{\partial^2 L}{\partial P_{g_i}\partial \lambda_{p_k}}\Delta \lambda_{p_k}$$
$$+ \sum_{k=NG+1}^{N}\frac{\partial^2 L}{\partial P_{g_i}\partial \lambda_{q_k}}\Delta \lambda_{q_k} + \sum_{k=NG+1}^{N}\frac{\partial^2 L}{\partial P_{g_i}\partial |V_k|}\Delta |V_k| = -\frac{\partial L}{\partial P_{g_i}}$$
$$\text{for } i = 1, 2, 3, \ldots, NG \quad (6.29a)$$

$$\sum_{k=1}^{NG}\frac{\partial^2 L}{\partial \delta_i\partial P_{g_k}}\Delta P_{g_k} + \sum_{k=2}^{N}\frac{\partial^2 L}{\partial \delta_i\partial \delta_k}\Delta \delta_k + \sum_{k=1}^{NG}\frac{\partial^2 L}{\partial \delta_i\partial \lambda_{p_k}}\Delta \lambda_{p_k}$$
$$+ \sum_{k=NG+1}^{N}\frac{\partial^2 L}{\partial \delta_i\partial \lambda_{q_k}}\Delta \lambda_{q_k} + \sum_{k=NG+1}^{N}\frac{\partial^2 L}{\partial \delta_i\partial |V_k|}\Delta |V_k| = -\frac{\partial L}{\partial \delta_i}$$
$$\text{for } i = 2, 3, \ldots, N \quad (6.29b)$$

$$\sum_{k=1}^{NG}\frac{\partial^2 L}{\partial \lambda_{p_i}\partial P_{g_k}}\Delta P_{g_k} + \sum_{k=2}^{N}\frac{\partial^2 L}{\partial \lambda_{p_i}\partial \delta_k}\Delta \delta_k + \sum_{k=1}^{NG}\frac{\partial^2 L}{\lambda_{p_i}\partial \lambda_{p_k}}\Delta \lambda_{p_k}$$
$$+ \sum_{k=NG+1}^{N}\frac{\partial^2 L}{\partial \lambda_{p_i}\partial \lambda_{q_k}}\Delta \lambda_{q_k} + \sum_{k=NG+1}^{N}\frac{\partial^2 L}{\partial \lambda_{p_i}\partial |V_k|}\Delta |V_k| = -\frac{\partial L}{\partial \lambda_{p_i}}$$
$$\text{for } i = 1, 2, 3, \ldots, N \quad (6.29c)$$

$$\sum_{k=1}^{NG}\frac{\partial^2 L}{\partial |V_i|\partial P_{g_k}}\Delta P_{g_k} + \sum_{k=2}^{N}\frac{\partial^2 L}{\partial |V_i|\partial \delta_k}\Delta \delta_k + \sum_{k=1}^{NG}\frac{\partial^2 L}{|V_i|\partial \lambda_{p_k}}\Delta \lambda_{p_k}$$
$$+ \sum_{k=NG+1}^{N}\frac{\partial^2 L}{\partial |V_i|\partial \lambda_{q_k}}\Delta \lambda_{q_k} + \sum_{k=NG+1}^{N}\frac{\partial^2 L}{\partial |V_i|\partial |V_k|}\Delta |V_k| = -\frac{\partial L}{\partial |V_i|}$$
$$\text{for } i = (NG+1), \ldots, N \quad (6.29d)$$

Assessment of Voltage Stability and Security 171

$$\sum_{k=1}^{NG} \frac{\partial^2 L}{\partial \lambda_{q_i} \partial P_{g_k}} \Delta P_{g_k} + \sum_{k=2}^{N} \frac{\partial^2 L}{\partial \lambda_{q_i} \partial \delta_k} \Delta \delta_k + \sum_{k=1}^{NG} \frac{\partial^2 L}{\lambda_{q_i} \partial \lambda_{p_k}} \Delta \lambda_{p_k}$$

$$+ \sum_{k=NG+1}^{N} \frac{\partial^2 L}{\partial \lambda_{q_i} \partial \lambda_{q_k}} \Delta \lambda_{q_k} + \sum_{k=NG+1}^{N} \frac{\partial^2 L}{\partial \lambda_{q_i} \partial |V_k|} \Delta |V_k| = -\frac{\partial L}{\partial \lambda_{q_i}}$$

$$\text{for } i = (NG+1), \ldots, N \quad (6.29e)$$

Eqs. (6.29a–6.29e) can be written in matrix form as shown in Eq. (6.30)

$$\begin{bmatrix} \frac{\partial^2 L}{\partial P_{g_i} \partial P_{g_k}} & 0 & \frac{\partial^2 L}{\partial P_{g_i} \partial \lambda_{p_k}} & 0 & 0 \\ 0 & \frac{\partial^2 L}{\partial \delta_i \partial \delta_k} & \frac{\partial^2 L}{\partial \delta_i \partial \lambda_{p_k}} & \frac{\partial^2 L}{\partial \delta_i \partial \lambda_{q_k}} & \frac{\partial^2 L}{\partial \delta_i \partial |V_k|} \\ \frac{\partial^2 L}{\partial \lambda_{p_i} \partial P_{g_k}} & \frac{\partial^2 L}{\partial P_{g_i} \partial P_{g_k}} & 0 & 0 & \frac{\partial^2 L}{\partial \lambda_{p_i} \partial |V_k|} \\ 0 & \frac{\partial^2 L}{\partial |V_i| \partial \delta_k} & \frac{\partial^2 L}{\partial P|V_i| \partial \lambda_{p_k}} & \frac{\partial^2 L}{\partial P|V_i| \partial \lambda_{q_k}} & \frac{\partial^2 L}{\partial |V_i| \partial |V_k|} \\ 0 & \frac{\partial^2 L}{\partial \lambda_{p_i} \partial \delta_k} & 0 & 0 & \frac{\partial^2 L}{\partial \lambda_{p_i} \partial |V_k|} \end{bmatrix} \begin{bmatrix} \Delta P_{g_i} \\ \Delta \delta_i \\ \Delta \lambda_{p_i} \\ \Delta \lambda_{q_i} \\ \Delta |V_i| \end{bmatrix} = \begin{bmatrix} -\frac{\partial L}{\partial P_{g_i}} \\ -\frac{\partial L}{\partial \delta_i} \\ -\frac{\partial L}{\partial \lambda_{p_i}} \\ -\frac{\partial L}{\partial |V_i|} \\ -\frac{\partial L}{\partial \lambda_{q_i}} \end{bmatrix} \quad (6.30a)$$

$$\underbrace{\begin{bmatrix} H_{P_g P_g} & 0 & H_{P_g \lambda_p} & 0 & 0 \\ 0 & H_{\delta\delta} & H_{\delta\lambda_p} & H_{\delta\lambda_q} & H_{\delta|V|} \\ H_{\lambda_g P_g} & H_{\lambda_p \delta} & 0 & 0 & H_{\lambda_p|V|} \\ 0 & H_{|V|\delta} & H_{|V|\lambda_p} & H_{|V|\lambda_q} & H_{|V||V|} \\ 0 & H_{\lambda_q \delta} & 0 & 0 & H_{\lambda_q |V|} \end{bmatrix}}_{[H]} \begin{bmatrix} \Delta P_g \\ \Delta \delta \\ \Delta \lambda_p \\ \Delta \lambda_q \\ \Delta |V| \end{bmatrix} = \underbrace{\begin{bmatrix} J_{P_g} \\ J_{\delta} \\ J_{\lambda_p} \\ J_{|V|} \\ J_{\lambda_q} \end{bmatrix}}_{[J]}$$

(6.30b)

or $\qquad$ [H] [Change in control variables] = [J] $\qquad$ (6.30c)

Here [H] and [J] are conventional *Hessian* and *Jacobian* matrices respectively. Starting from initial data of an interconnected power system, optimal power flow solution may be achieved by solving Eq. (6.30) using the iterative process. In this method, there is no need for a separate load flow study; moreover this method has been observed to be faster with more accuracy as in optimal power flow both power generations and voltages are calculated simultaneously considering the economics of operation with all system constraints taken into account.

Again, $\partial Q_i/\partial |V_i|$ being the reactive power sensitivity of the *i*th bus, it inherently indicates the degree of weakness for the *i*th bus (as with $\partial Q_i/\partial |V_i|$ being high, $\partial |V_i|/\partial Q_i$ becomes low indicating minimum change in $|V_i|$ for variation in Q-status of the bus). Thus in a multibus power network, at bus-*i*, $\partial Q_i/\partial |V_i|$ being higher, the degree of weakness of the *i*th bus becomes lesser. $\partial Q_i/\partial |V_i|$ can be calculated here and comparing the magnitudes of all $\partial Q_i/\partial |V_i|$ the weakest bus in the system can be identified. Also the determinant of Hessian matrix ([H] matrix in Eq. (6.30c)) approaching zero, there will be a large change in control parameters of this state and therefore the critical point of optimal power flow solution, beyond which there will be no feasible value of generation, can be identified. This state also represents the voltage stability limit.

The reactive power sensitivity of any load bus ($dQ_i/d|V_i|$) can easily be computed from optimal power flow algorithm (optimising cost functions governed by real and reactive power constraints) as explained above. With higher reactive power sensitivity, the change in bus voltage will be lower indicating a lower degree of bus sensitivity. The degree of bus sensitivity can be defined as

$$\text{Degree of bus sensitivity} = \frac{dQ_{i_{\text{old}}}}{d|V_i|_{\text{old}}} - \frac{dQ_{i_{\text{new}}}}{d|V_i|_{\text{new}}}$$

Thus, if there be any change in load status at any bus of the power system, the effect can easily be seen from the optimal power flow solution and as the system becomes heavily stressed, the voltage stability region is likely to be depleted as well as bus sensitivity is increased pushing the system towards voltage instability. The SVC model (FC-TCR) can be incorporated in optimal power flow algorithm in order to observe its effect in improving the voltage stable states in the load buses along with the improvement in the degree of bus sensitivity and system economics of the power system. The detailed flow chart of this proposed scheme is shown in Figure 6.11.

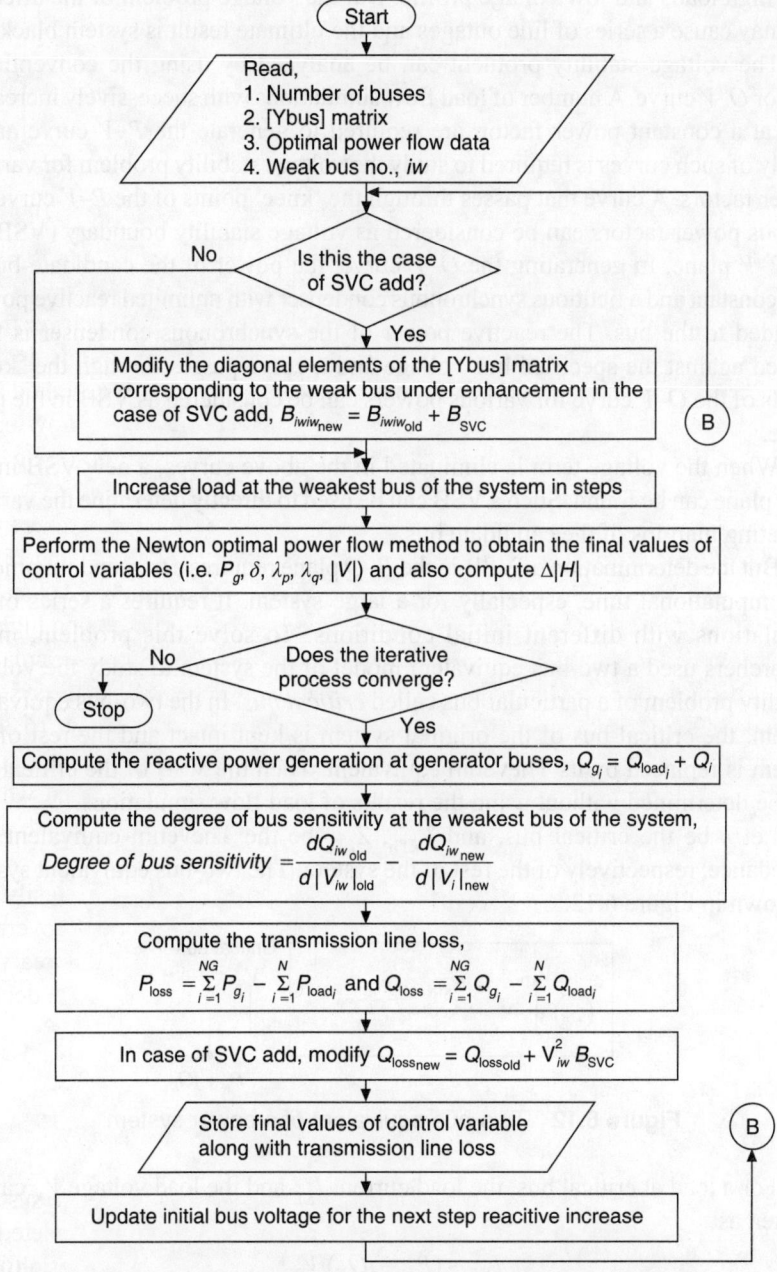

Figure 6.11 Flow chart of the method of determining voltage stability from OPF technique.

6.3.12 Method Using Stability Boundary in *P–Q* Plane

Voltage instability is a local phenomenon and it occurs at a bus within an area with high loads and low voltage profile. But the voltage problem of the affected bus may cause a series of line outages and the ultimate result is system blackout.

The voltage stability problem can be analyzed by using the conventional *P–V* or *Q–V* curve. A number of load flow simulations with successively increased load at a constant power factor, are required to generate the *P–V* curve and a family of such curves is required to study the voltage stability problem for various power factors. A curve that passes through the 'knee' points of the *P–V* curve for various power factors can be considered as voltage stability boundary (VSB) in the *P–V* plane. In generating the *Q–V* curve, the power of the candidate bus is kept constant and a fictitious synchronous condenser with unlimited reactive power, is added to the bus. The reactive power of the synchronous condenser is then plotted against the specified bus voltage and a curve passes through the 'knee' points of the *Q–V* curve for various powers can be considered as VSB in the *Q–V* plane.

When the voltage term is eliminated in the above curves, a new VSB in the *P–Q* plane can be found. Such a VSB can be used to directly determine the various operating margins of the candidate bus.

But the determination of VSB in the *P–Q* plane requires a tremendous amount of computational time, especially for a large system. It requires a series of LF simulations with different initial conditions. To solve this problem, many researchers used a two-bus equivalent model of the system to study the voltage stability problem of a particular bus called *critical bus*. In the two-bus equivalent system, the critical bus of the original system is kept intact and the rest of the system is replaced by its Thevenin equivalent. Then the VSB of the critical bus can be determined without using the results of load flow simulations.

Let k be the critical bus, and V_{kTh}, Z_{kTh} be the Thevenin equivalent and impedance, respectively of the rest of the system. The two-bus equivalent system is shown in Figure 6.12.

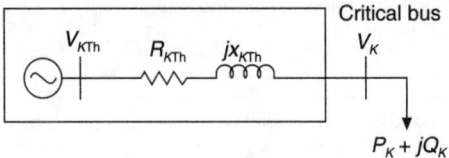

Figure 6.12 Two-bus equivalent of a power system.

For a load at critical bus, the load current I_{KL} and the load voltage V_K can be written as

$$I_{KL} = (P_K - jQ_K)/V_K^* \qquad (6.31)$$

$$V_K = V_{KTh} - Z_{KTh} I_{KL} \qquad (6.32)$$

After calculations from the above two equations, the load voltage V_K of the critical bus can be written as

$$V_K^4 + 2(R_{K\text{Th}}P_K + X_{K\text{Th}}Q_K)V_K^2 - V_{K\text{Th}}^2 V_K^2 + (R_{K\text{Th}}^2 + X_{K\text{Th}}^2)(P_K^2 + Q_K^2) = 0 \tag{6.33}$$

Equation (6.33) has four possible solutions. However, under normal operating conditions, the equation has only two feasible solutions. The higher voltage V_K^H is known as the stable solution and the lower voltage V_K^L is called the unstable solution. The difference between these two solutions may be considered as an index to measure the voltage stability margin. As the load increases, the stable solution decreases while the unstable solution increases. At the voltage collapse point, both the solutions become same. Thus, at the voltage collapse point,

$$V_K^H = V_K^L \tag{6.34}$$

Expressions of V_K^H and V_K^L for a feasible load condition, can be obtained from Eq. (6.33). With these expressions, Eq. (6.34) can be rewritten as

$$f(P_K, Q_K, V_{K\text{Th}}, R_{K\text{Th}}, X_{K\text{Th}}) = 0 \tag{6.35}$$

for a given load bus K, $V_{K\text{Th}}$, $R_{K\text{Th}}$, $X_{K\text{Th}}$ are fixed. When the above quantities are constant, the VSB of the load bus in the P–Q plane can easily be generated from the solution of Eq. (6.35).

The set of loads that satisfies Eq. (6.35) can be considered as critical load $(P_K^{\text{cr}}, Q_K^{\text{cr}})$. A typical variation of Q_K^{cr} against P_K^{cr} is considered as VSB in the P–Q plane (Figure 6.13).

The shaded area represents the infeasible operating region and the unshaded area represents the feasible operating region. The boundary between the feasible and the infeasible regions is called VSB.

For a given initial operating point x in the feasible operating region, distances xx_1 and xx_2 represent the active and reactive power margins of the critical bus. Similarly, the distance xx_3 represents the apparent power margin at a power factor angle 'a'.

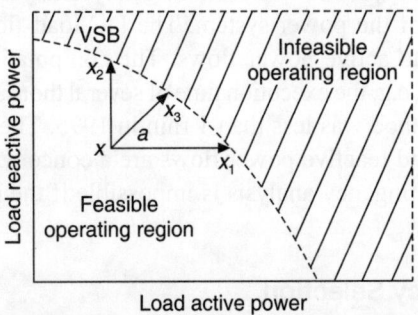

Figure 6.13 VSB in the P–Q plane.

6.4 CONTINGENCY ANALYSIS

The result of the contingency analysis is the classification of the power system into secure and insecure states. This is an essential part of the security analysis. The contingency analysis programs are based on a model of the power system and are used to study outages and notify the operators of any potential overloads or out-of-limit voltages. Contingency analysis is a time-consuming process when the number of contingencies is large. The contingency analysis of one thousand outages would take about 16 min, if one outage case were studied in 1 second. This would be useful if the power system conditions did not change over that period of time. Most of the time spent running contingency analysis would go for solutions of the load-flow model that discovers that there are no problems. The on-line contingency analysis is usually performed with a short contingency list using simplified computation methods. The contingency list is selected according to contingency ranking.

6.4.1 Determination of Power System Post-disturbance State

Contingencies arise due to scheduled outage, component switching in order to optimise power system operation, or unscheduled outage due to a fault. The contingency analysis of scheduled outage is based on a predicted operation point. Component switching in order to optimise power system operation requires contingency analysis of the current operation point. The contingency analysis of unscheduled outages is done automatically in the on-line security assessment.

The contingencies are classified as transmission component (lines, transformers or substation buses) and generator contingencies. The effect of contingency is seen as a change in power flows and voltages in the power system. The outage of a generation unit also increases the production of other generators due to shortage of power production. The modelling of generator contingency requires distribution of generated power to all generators taking part in frequency control. Similarly, additional losses caused by contingency should be distributed among generators.

To speed up the execution, the contingency analysis procedure uses an approximate model of the power system. The DC load-flow provides accurate results with respect to active power flows. The computation time of DC load-flow is extremely fast, e.g. the execution time of several thousand outages computed by DC load-flow method was less than 1 rnin in 1995. The full AC load-flow is required if voltage and reactive power flows are a concern. However, the use of AC load-flow in contingency analysis is impossible if the number of outages is several thousands.

6.4.2 Contingency Selection

The selection of contingencies is needed to reduce the computation time of contingency analysis. Operators know from experience which potential outages will cause trouble. The danger is that the past experience may not be sufficient under the changing network conditions. There is a possibility that one of the cases they have assumed to be safe may in fact present a problem because some of the operators' assumptions used in making the list are no longer correct.

The contingency list is dependent on the power system operation point; thus it must be periodically updated. Another way to reduce the list to be studied is to calculate the list as often as the contingency analysis itself is done. To do this requires a fast approximate evaluation to discover those outage cases which might present a problem and require further detailed evaluation. Many methods have been proposed to solve this problem. The selection of contingencies is based on contingency ranking methods, and contingency ranking is used to estimate the criticality of studied outages. Contingency ranking methods assemble the appropriate severity indices using the individual monitored quantities such as voltages, branch flows and reactive generation.

Contingency list

One way of building the contingency list is based on the fast load-flow solution (usually an approximate one) and ranking the contingencies according to its results. The DC load-flow is commonly used for contingency ranking which is a completely linear, non-iterative, and load-flow algorithm. This method is used to eliminate most of the cases and to run AC load-flow analysis on the critical cases.

The DC load-flow solution may be further speeded up by the bounding methods. These methods determine the parts of the network in which the active power flow limit violations may occur and a linear incremental solution (by using linear sensitivity factors, i.e. generation shift and line outage distribution factors) is performed only for the selected areas. The solution of AC load-flow may be speeded up by attempts to localise the outage effects and to speed up the load-flow solution. The efficiency of the load-flow solution has been improved by means of approximate or partial solutions (such as by following the below described 1P1Q method) or by using network equivalents. The bounding method may also be applied to the AC load-flow solution. Another localised method for AC load-flow solution is the zero mismatch method.

1P1Q method

A commonly used contingency ranking method is the 1P1Q method where a decoupled load-flow is used to give approximate values for line flows and voltages. The name of the method comes from the solution procedure, where the decoupled load-flow is interrupted after one iteration of active power/voltage angle and reactive power/voltage calculations. A reasonable performance index is given from the solution of the first iteration of decoupled load-flow. The decoupled load-flow has the advantage that a matrix inversion can be incorporated into it to simulate the outage of lines without reinverting the system Jacobian matrix at each iteration. Compared to DC load-flow method, another advantage of decoupled load-flow is the fact that the solution of voltages is given.

The ranking of contingencies is based on the performance index used for estimating the state of the post-disturbance operation point. The performance index is typically a penalty function, which indicates the crossing of limit values of power system parameters, such as voltages and line flows. The index should have a high value for critical contingencies and a low value for non-critical contingencies. The advantage of performance indices is a clear and a simple analysis, the disadvantage being, e.g. an unclear setting of limit values for line flows and voltages in the voltage stability analysis. An example of performance index is given in Eq. (6.36).

$$PI = \sum_{\substack{\text{all lines} \\ j}} \left(\frac{P_j}{P_j^{\max}}\right)^{2n} + \sum_{\substack{\text{all buses} \\ i}} \left(\frac{\Delta |E_i|}{\Delta |E|^{\max}}\right)^{2m} \quad (6.36)$$

where P_j is the line flow at the solution of 1P1Q procedure, $P_j^{\max}$ is the maximum value of line flow, n and m are high numbers, $\Delta |E_i|$ is the difference between the voltage at the solution of the 1P1Q procedure and the base case voltage, and $\Delta |E|^{\max}$ is the value set by operators indicating how much they want to limit voltage changes on the outage.

Consideration of power system non-linearity

The progress of computer and software technology has made it possible to analyse bigger and more demanding studies. A new topic in contingency analysis is the consideration of non-linear functioning of the power system. Voltage stability analysis requires new kinds of contingency analysis methods due to the non-linearity of the problem. The contingency analysis of voltage stability studies is even more demanding than the analysis of single AC load-flow due to the non-linearity of the problem.

Static contingency ranking is traditionally based on the properties of pre and post-disturbance operation points and aims at quickly identifying the contingencies that will lead to some operating limit violation. However, the ranking of contingencies might change due to power system non-linearity when the power system is operated temporarily close to security limits. The objective of voltage stability contingency ranking is ideally to identify the contingency that yields the lowest margin. The ranking should take into account both the disturbance and the direction of system stress. Accurate contingency ranking requires the computation of voltage collapse points for each contingency. The ranking of contingencies is then based on post-disturbance voltage stability margins. The computation of voltage stability margin for each contingency requires a large amount of computation and is not a practical solution to the problem.

Figure 6.14 presents PV-curves for one pre-disturbance case and two post-disturbance cases. Sometimes post-disturbance PV-curves intersect at some parameter value due to power system non-linearity. A typical reason for PV-curve intersection is the action of generator reactive power limit, which causes a strong decrement in load voltages. In practice, it is almost impossible to approximate

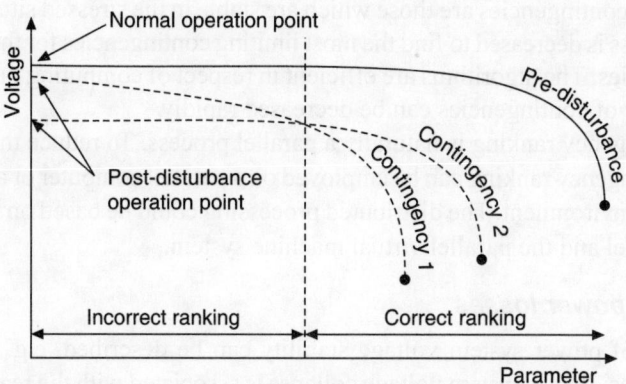

Figure 6.14 Intersection of post-disturbance *PV*-curves.

contingency ranking correctly based on pre- or post-disturbance operation points, if those are far from the voltage collapse point. The contingency ranking is different when seen from the operation point than from the stressed point.

6.4.3 Contingency Ranking Methods for Voltage Stability Studies

The contingency ranking methods for voltage stability analysis approximate the true contingency ranking without the computation of voltage stability margins. If the contingency ranking is done close to the most critical voltage collapse point (at the stressed operation point), then the ranking can be done more reliably than that at the operation point. The methods are based on full load-flow computation for every contingency at a stressed operation point. The methods are reactive power losses, minimum singular value of load-flow Jacobian matrix, curve fitting technique to approximate voltage stability margin and clustering of contingencies to critical and non-critical contingencies.

Other contingency ranking methods for voltage stability analysis are based on sensitivities of voltage stability margin, simultaneous computation of multiple contingency cases and parallel/distributed computation algorithms. The sensitivity method computes a *PV*-curve to obtain a pre-disturbance voltage stability margin. Then linear sensitivities of the margin for each contingency are computed. These sensitivities are then used to estimate the post-disturbance voltage stability margins. The advantage of the method is fast computation compared to other methods. The accuracy of the method is good and can be improved by using quadratic sensitivities. Problems may emerge when contingencies shift the voltage critical area or change the set of generators that are at the reactive power limits at the voltage collapse point.

There are some proposals to calculate multiple contingency cases simultaneously to rank contingencies according to voltage stability. These iterative methods compute load-flows or perform dynamic simulations to determine the stability of contingency cases in stressed situations. The idea of these methods is to eliminate as many contingency cases as possible at the first iteration. The eliminated contingencies are those which are stable in the stressed situation. Then system stress is decreased to find the most limiting contingencies for the remaining contingencies. The algorithms are efficient in respect of computing time, because the number of contingencies can be decreased rapidly.

Contingency ranking is naturally a parallel process. To reduce the execution time, contingency ranking can be employed on a parallel computer or a distributed computer environment. The distributed processing could be based on the client or server model and the parallel virtual machine system.

Reactive power losses

The state of power system voltage stability can be described, e.g. by reactive power losses. Power system voltage collapse is associated with the reactive power demands not being met because of limitations in the production and transmission of reactive power. When the power system is stressed, reactive power losses increase compared to the operation point. There is a performance index that measures reactive power loss in a selected area to rank line outages. The performance index approximates the post-disturbance reactive power losses according to the pre-disturbance load-flow solution. This performance index can be improved by computing the solutions of post-disturbance load-flows. In that case, the reactive power losses of outages need not be estimated and the ranking of contingencies can be directly based on them. It is assumed that reactive power losses will be high for critical contingencies and low for non-critical contingencies compared to the pre-disturbance value in the proposed contingency ranking method

Minimum singular value

The minimum singular value of the load-flow Jacobian matrix is zero at the voltage collapse point. It is used as an indicator to quantify proximity to post-disturbance maximum loading point. The use of the indicator requires the computation of post-disturbance load-flows for each outage. Although singular value decomposition is a linear description of operation point, it is assumed to give an accurate enough contingency ranking for voltage stability studies. The intersection of contingency case PV-curves cannot be realised with this method. The value of the minimum singular value of the load-flow Jacobian matrix is also sensitive to limitations and changes in reactive power output. Computing the minimum singular values at the stressed operation point can increase the accuracy of the method.

Curve fitting method

The idea for voltage stability margin approximation with curve fitting technique is taken from several works in the literature. This method needs at least three stable post-disturbance load-flows to approximate the maximum loading point. Computed values are used to fit second order polynomials for chosen power system nodes. The second order polynomial is readily applicable to voltage stability margin approximation because the shape of PV-curve is parabolic.

The accuracy of the voltage stability margin approximated by the curve fitting technique is fully dependent on the fitting points. If all the fitting points are located at the flat part of the PV-curve, the estimates of critical state variables and critical parameters are sensitive to small changes in fitting points. The voltage stability margin is not sensitive to the estimation errors if the number of power system nodes studied is large enough. It is also possible to remove the critical state variables and parameters which are clearly impossible from the computation of voltage stability margin. When the fitting points are located far from each other, it is possible to get a result where one of the fitting points is at the lower side of the PV-curve. This is a result of incorrect curve fitting, because the fitting points used in the curve fitting should be located at the upper side of the PV-curve. Curve fitting is not explicit, which causes these problems.

The selection of step length in the computation of operation points (using e.g. the maximum loading point method) is critical for the accuracy of voltage stability margin approximation. Computed operation points should not be too far away from each other, otherwise approximation of the maximum loading point may lead to strange results. On the other hand, step length should be large enough to avoid constant voltages in fitting points.

The accuracy of voltage stability margin approximation can also be improved by adding fitting points. The second order polynomial goes through the fitting points exactly if the number of points is three. If the number of fitting points is larger than three, the curve fit is done in the sense of least square error and the polynomial does not necessarily go through the fitting points.

Contingency clustering

Most pattern recognition methods proposed for contingency analysis are based on classification methods. The application of classification to contingency ranking requires off-line training of the classifier with a large number of contingency studies. Clustering is distinct from classification methods and tries to find similarities between cases.

6.4.4 Comparison of Contingency Ranking Methods

The contingency ranking methods described above have different characteristics. The curve fitting technique is the most accurate method, but it also requires a lot of computation compared to other methods. It provides a good approximation for a voltage stability margin in order to rank contingencies. This method can also be used for a first approximation of security analysis in the on-line mode. In the on-line mode the security analysis is usually realised in parallel form, where all contingencies are studied almost simultaneously. When there are three load-flow solutions available for all contingency cases, the curve fitting technique can be applied to approximate voltage stability margins. The first approximation of voltage stability margins for different contingency cases provides an opportunity to concentrate the computation capacity on the most critical contingencies and eliminate the others. The curve fitting method is the most recommended method of the studied methods due to its accuracy and capability to approximate the voltage stability margin.

Methods based on reactive power losses and minimum singular value almost rank equal when considering the accuracy of contingency ranking. The computation of reactive power losses is naturally faster than the computation of the minimum singular value of load-flow Jacobian matrix. Both of these methods are intuitively easy to adapt, because they offer clear physical or mathematical explanations concering power system voltage stability. They are easy to implement compared to existing analysis tools. There are also other similar indices like reactive power reserves which can be combined with the above methods to achieve even better results.

The contingency clustering method is not capable of ranking contingencies. It is a valuable method when a large amount of data of contingency studies is to be analysed. The method provides a way to analyse both the variables and the cases. The purpose of the proposed method is to find similarities in contingency cases in order to find the most critical contingency. The value of the proposed method is clear when the data includes cases in a wide range of operation points. The data needed in contingency clustering comes from the planning studies and the computation of on-line transfer capacities. When transmission companies start to apply risk management or uncertainty handling in transmission network planning and operation, there are a huge amount of stability studies which must be analysed in a systematic way.

EXERCISES

1. Explain with the help of a block diagram the different power system states.
2. What is security assessment? Describe an off-line security assessment procedure for a multi-bus system.
3. What is transfer capacity? What do you mean by voltage stability margin?
4. Discuss the utility of the following two methods of assessing voltage collapse:
   (i) method of minimum singular value
   (ii) point of collapse method.
5. How do you find the point of voltage collapse using continuation load flow?
6. How would you compute the maximum loading point in a power system?
7. How do you define V–Q sensitivity of a network using eigenvalue analysis?
8. Develop a bus equivalencing technique to assess the stability of the power system with respect to voltage collapse.
9. Develop the concept of voltage stability using the method of optimal power flow.
10. Obtain voltage stability boundary in P–Q plane.
11. Discuss how a contingency can affect the voltage stable state in a power system. How can you assess it?
12. Discuss a few contingency ranking methods in voltage stability studies.

Chapter 7

Voltage Control and Improvement of Voltage Stability in Power Transmission Systems

7.1 INTRODUCTION

The analyses and discussions in the preceeding chapters clearly reveal that the transmission system is very susceptible to the real and reactive power changes. Frequency deviation being the result of real power mismatch between the generation and demand, voltage mismatch is the sole indicator of reactive power imbalance in the system. For medium and long EHV AC transmission systems, the system voltage at the receiving-end bus of an uncompensated line may drop well below the acceptable limit, in case of power transfer corresponding to line ratings. It has been described how the voltage instability may occur for these systems, particularly when the nodal short-circuit strength is low and the generating capacity is comparable with the loads connected at the load bus. Outages play a leading role in driving the system to voltage unstable zones which may ultimately lead to voltage collapse. The reactive power transfer constraints, lack of local reactive power injecting devices, radial system configuration, etc. also adversely affect the operation of the system at a stable voltage state.

However, in transmission systems, where the capacity of generation is comparable with that of the loads at the load bus and when the power is to be shipped through long lines, the voltage state operates in the proximity of voltage collapse and any reactive power disturbance may push the entire system to the verge of voltage instability. The operator at the load centre may not see any appreciable change in load current and load bus voltage profile till the system enters into an unstable voltage state. Lack of reactive power reserve, a poor co-ordination and failure of the communication network may render the operator helpless and voltage collapse at the load bus may take place. The operator, in a bid to save the load bus secondary voltage, may go for uncontrolled tap changing which though temporarily improves the distribution or subtransmission voltage, does not improve the transmission voltage. It has been indicated earlier how the system voltage may collapse under this condition. However, till the system voltage enters into the unstable zone, tap-changer operation may improve the voltage at the load bus provided the system has not lost its reactive power reserve. The following discussions are presented to highlight the merits and demerits of an on-load tap-changer for the control of system voltage in a transmission line.

7.2 ROLE OF TRANSFORMERS IN VOLTAGE CONTROL OF A POWER SYSTEM

By changing the transformation ratio, the voltage in the secondary side of any bus can be varied. Power transformers, being used extensively for the control of transmission and subtransmission voltage of a network utilising this principle, may be either manual or automatic. The latter, usually called On Load Tap Changers (OLTC), are usually arranged to regulate the bus voltage in order to keep the operating voltage of the regulated bus within some acceptable limits. The following analysis is aimed at presenting an investigation for the best choice of the regulated bus. The best operating principle for the OLTC has also been discussed.

7.2.1 Modelling of Transformers

Figure 7.1 represents a two-winding transformer with a tap-changer, whose off-nominal tap has a turns ratio of 1 : a. connected to bus 1 of an EHV power system and has been represented by an equivalent circuit in Figure 7.2. The short-circuit impedance being purely reactive, the shunt admittances Y_{sh1} and Y_{sh2} are given by

$$Y_{sh1} = a(a-1)Y \tag{7.1}$$

and $\qquad Y_{sh2} = (1-a)Y \tag{7.2}$

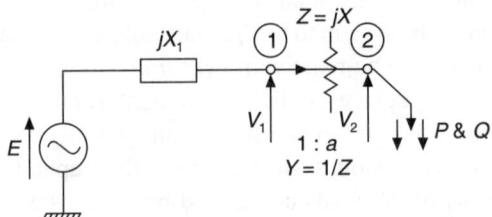

Figure 7.1 Schematic of a power transformer with off-nominal tap.

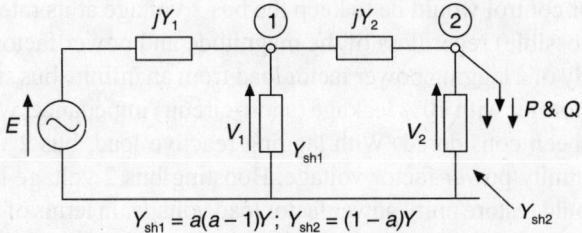

$Y_{sh1} = a(a-1)Y;\ Y_{sh2} = (1-a)Y$

Figure 7.2 Equivalent circuit of a power transformer with off-nominal tap.

It may be noted that if $a > 1$, Y_{sh_1} is inductive, Y_{sh2} is capacitive. But if $a = 1$, $Y_{sh1} = Y_{sh2} = 0$ and if $a < 1$, Y_{sh_1} is capacitive and Y_{sh2} is inductive.

Upon correlating the preceding statements with the off-nominal turns ratio 1 : a, it is evident that the equivalent circuit of the transformer always places a capacitor at the bus whose voltage is to be raised by the tap-changer and a reactor at the other bus. Furthermore, the reactive power produced/absorbed by this added capacitor/reactor combination is a function of 'a' as described by Eqs. (7.1) and (7.2). This reactive power generally increases, either quadratically or linearly for Y_{sh_1} and Y_{sh2} respectively as the size of the tap 'a' increases above unity or decreases below unity, as depicted in Table 7.1.

Table 7.1 Effect of 'a' on Y_{sh_1} and Y_{sh2}

a	Y_{sh_1}/Y	Y_{sh2}/Y
1.25	0.31	−0.25
1.20	0.24	−0.20
1.15	0.17	−0.15
1.10	0.11	−0.10
1.05	0.05	−0.05
1.00	0.00	0.00
0.95	−0.05	0.05
0.90	−0.09	0.10
0.85	−0.13	0.15
0.80	−0.16	0.20

7.2.2 Case I: Voltage Control of a Radial Load

Figure 7.2 represents the supply of a passive radial load at bus 2, through a power transformer with an automatic tap-changer connected between buses 1 and 2, from a source system connected to bus. In such a system, bus 1 would normally be a high voltage transmission level bus and bus 2 would be a low voltage utilisation bus. However, a voltage step-up from bus 1 to bus 2 would also be possible. This is rare since normally the load at bus 2 would be of a lagging power factor although, in a few cases, leading or unity power factor loads may also be possible. The strength of the source system at bus 1 could vary over a wide range from case to case. Control criteria for the voltage at bus 2 may vary but typically, the objective of tap-changer control would be to keep the bus 2 voltage at its rated value (or as near to it as possible) regardless of the magnitude and power factor of the load.

The supply of a lagging power factor load from an infinite bus supply through a power transformer with 10% leakage (short-circuit) impedance, which is purely reactive, has been considered. With lagging reactive load, bus 2 voltage would drop from its unity power factor voltage. Boosting bus 2 voltage by means of a tap change would restore unity power factor load voltage. In terms of the equivalent circuit of the tap-changer, the tap-changer achieves this voltage correction by adding a shunt capacitor (Y_{sh2} in Figure 7.2 and Table 7.1) at bus 2 to cancel the reactive-part of the load, thus restoring unity power factor branch (aY) loading. At the same time as almost equally rated shunt reactor (Y_{sh1} of Figure 7.2 and Table 7.1) is added at bus 1, effectively transferring the reactive part of the load from bus 2 to bus 1, and eliminating the voltage drop in the transformer caused by this reactive load. If a higher voltage boost is used instead of a 10% voltage boost in the foregoing analysis, the load at bus 2 would be over-corrected from lagging power factor to leading power factor. Then bus 2 would have over-voltage and the transformer would still be overloaded. Current in branch (aY) will then go through minima near unity power factor load, as the tap-changer varies its taps through a range from maximum buck to maximum boost. This is similar to the 'V' curves of a synchronous machine as its excitation is varied from under-excited to over-excited operation. This minimum branch loading condition can be considered to be an optimum operating strategy for the controller of the automatic tap-changer. In this case, it is clearly the bus 2 which should be the regulated bus as the voltage of bus 1 cannot vary.

In practice, the supply system at bus 1 of Figure 7.2 will not be an infinite bus but will have a finite short-circuit capacity and the foregoing effects will be modified to some extent. The shunt capacitor will, however, be identical to what it was before, since it depends only on the short-circuit impedance of the transformer and the tap-changer turns ratio. Any difference between the reactive power in Y_{sh1} and Y_{sh2} will now however be supplied from the finite system behind bus 1 and there will therefore be some additional voltage regulation at both buses 1 and 2. This is due to the difference between the reactive power of Y_{sh1} and Y_{sh2} that must be supplied through impedance jX_1 (which is now a non-zero quantity).

7.2.3 Case II: Tie-Transformer between Systems of Various Strengths

Figure 7.3 depicts a power transformer with an automatic tap-changer when used as a 'tie-transformer' between two different systems, or between two different transmission levels within a system. The load is shown connected to bus 2. The strength or short-circuit capacity of the two systems varies in the three cases considered. These systems will be considered individually to reveal their differences.

Voltage Control & Improvement of Voltage Stability in Power Transmission Systems   187

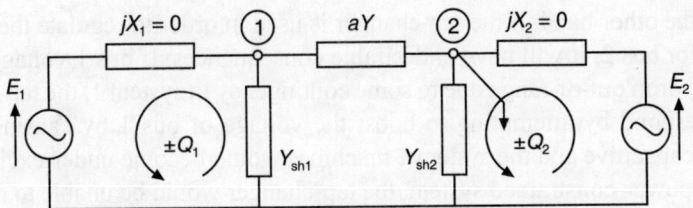

Figure 7.3(a) Operating states of tie-transformers.

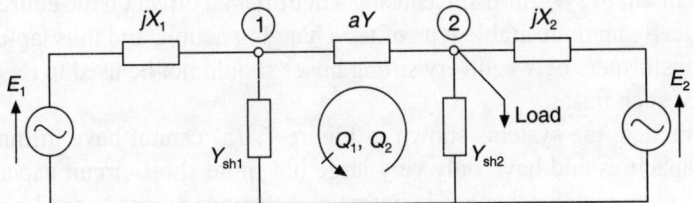

Figure 7.3(b) Operating states of tie-transformers

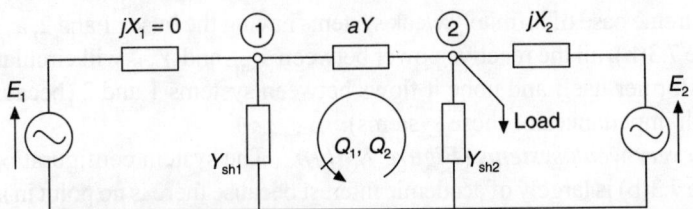

Figure 7.3(c) Operating states of tie-transformers.

Two very strong systems (Figure 7.3(a)): If the systems at the back of buses 1 and 2 are both very strong or infinite in the limit as depicted in Figure 7.3(a) then bus voltages V_1 and V_2 cannot vary. Any off-nominal turns ratios achieved by tap-changer action will only result in reactive power circulation by the loops for Q_1 and Q_2. The direction of the reactive power circulation would depend on which one of Y_{sh1} and Y_{sh2} is a source and which one a sink for reactive power. This in turn depends on whether $a > 1$ or $a < 1$. Reactive power between Y_{sh1} and Y_{sh2} cannot flow through the branch aY itself because such a reactive power flow would require a voltage change and the potential difference between buses 1 and 2 cannot change with infinite buses at buses 1 and 2. It may be noted that such reactive power circulation depends on the reactive power capabilities of systems 1 and 2.

Reactive power transfer from system 1 to system 2, or vice versa, as depicted in Figure 7.3(a), could easily be achieved by tie-transformer tap-changer control, much in the same manner as phase-shifting transformers achieve real power transfer between such systems. The amount of reactive power-transferable between systems 1 and 2 in this manner can be determined from the values of Y_{sh1} and Y_{sh2} as summarised in Table 7.1. This type of reactive power transfer can be regarded as a form of system voltage control.

On the other hand, if the tap-changer is used in order to regulate the voltage of bus 1 or bus 2, it will have undesirable consequences. If bus 1 voltage was to suddenly drop out-of-range due to some contingency in system 1, the tap-changer would respond by attempting to boost the voltage of bus 1 by causing Y_{sh_1} to become capacitive and the system 1 machines would become under-excited. In a reactive power-constrained system, the tap-changer would be unable to raise bus 1 voltage and it would continue changing taps to the extreme boost position thus making the system 1 machines severely under-excited. This may possibly cause them to fall out of synchronism, causing a detrimental effect on the entire system. This is clearly an undesirable type of tap-changing action, and thus tap-changers on tie-transformers between very strong buses should not be used to regulate the voltage of such buses.

In practice, the systems shown in Figure 7.3(a) cannot have infinite short-circuit capacities and have only very large but finite short-circuit capacities. In this case, there may be changes in potential difference across buses 1 and 2, and some reactive power circulation will take place between Y_{sh_1} and Y_{sh_2} through the transformer in addition to the reactive power flow between systems 1 and 2. In the extreme case of infinitely weak systems having the buses 1 and 2, as depicted in Figure 7.3(b), all the reactive power between Y_{sh_1} and Y_{sh_2} will circulate within the transformer itself and none it flows between systems 1 and 2 (because of the very high impedances of these systems).

Two very weak systems (Figure 7.3(b)): The system configuration shown in Figure 7.3(b) is largely of academic interest because there is no point in installing a 'tie-transformer' between two systems which are incapable, because of their very high impedances, of transferring any load between themselves. Figure 7.3(b) reveals the reactive power flow loop between Y_{sh_1} and Y_{sh_2} through the transformer itself.

One very strong, one very weak system (Figure 7.3(c)): With X_2 finite, however, there is possibility of using the tie-transformer to transfer load between systems 1 and 2 in addition to supplying radial loads at bus 2. Theoretically it is possible to transfer load in both directions but it is probably always from the stronger to the weaker system (bus 1 to bus 2), except perhaps under certain exceptional circumstances or emergencies. Under normal circumstances, therefore, with power transfer from system 1 to system 2, in addition to load at bus 2, the configuration in Figure 7.3(c) is effectively the same as that in Case I. Except that now it is the total load at bus 2, plus the power transfer to system whose power factor must be compensated by the tap-changer in order to minimise the branch aY current. The above discussion reveals that transformers can be used effectively for the voltage control of a receiving-end bus in a radial power transmission system where the source is comparatively much stronger than the load bus. Voltage control utilising transformers in the lines in-between two very strong or two very weak systems does not yield any effective result.

7.3 QUANTITATIVE METHOD TO DETERMINE THE TAP SETTING FOR VOLTAGE CONTROL USING OLTC AT A LOAD BUS

As the reactive power status at any bus governs the voltage, it is very important to determine the reactive power position at the receiving bus.

7.3.1 Reactive Power at Receiving Bus

In a steady operating state, the voltage at the receiving bus can be controlled by the injection of reactive power of proper type into the network. This may be achieved by the use of tap-changing trarsformers which redistribute the reactive power flow in order to boost the voltage with which they are connected. Here expressions have been summarised to determine the reactive power requirement for suitable voltage control. Expressions have also been shown to quantify the tap setting for steady state voltage control in an EHV power transmission system.

Representation of a four-terminal power system in terms of A, B, C, D constants gives the expression for the receiving-end reactive power as

$$Q_R = \frac{-C_2 + \sqrt{C_2^2 - 4 C_2 C_3}}{2 C_1} \tag{7.3}$$

where

and

$$\left. \begin{array}{l} C_1 = \dfrac{B^2}{V^2}, \ C_2 = -2AB \sin(\alpha - \beta) \\[2mm] C_3 = A^2 V^2 + \dfrac{B^2 P_R^2}{V^2} + 2ABP_R \cos(\alpha - \beta) - E^2 \end{array} \right\} \tag{7.4}$$

Equation (7.3) represents the receiving-end reactive power requirement.

7.3.2 Method of Voltage Control by Tap-Changing Transformers

It has been already indicated that by changing the transformation ratio, the voltage of the secondary circuit can be varied and thus voltage control obtained. This constitutes the most popular and wide-spread form of voltage control at all voltage levels. The method has been illustrated here for a radial transmission system with two tap-changing transformers in the equivalent, single phase network of Figure 7.4. It is required to determine the tap changing ratios needed to completely or partially compensate the voltage drop in the line in order to achieve desired voltage control at the receiving-end load terminals.

From Figure 7.4,

$$t_s E_1 = A(t_r E_2) + BI \tag{7.5}$$

where E_1 and E_2 are the nominal voltages at the ends of the line, the actual voltages being $t_s E_1$ and $t_r E_2$.

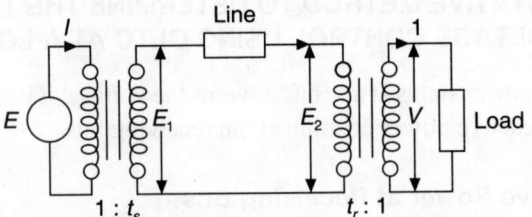

Figure 7.4 Voltage control scheme of tap-changing transformer
t_s—Sending-end tap ratio
t_r—Receiving-end tap ratio.

The receiving-end current is given by

$$I_R = \frac{P}{V} - j\left(\frac{P}{V}\right)\tan\phi \qquad (7.6)$$

where ϕ is positive for lagging power factor. The line current is given by

$$I = \frac{I_R V}{t_r E_2} = \left(\frac{P}{t_r E_2}\right)(1 - \tan\phi) \qquad (7.7)$$

Simplifying Eq. (7.5) and utilising Eq. (7.7),

$$t_s E_1 = A t_r E_2 + B\left(\frac{P}{t_r E_2}\right)(1 - j\tan\phi)$$

$$= A t_r E_2 (\cos\alpha + j\sin\alpha) + B(\cos\beta + j\sin\beta)\left(\frac{P}{t_r E_2}\right)(1 - j\tan\phi)$$

$$= \left\{ A t_r E_2 \cos\alpha + B\cos\beta\left(\frac{P}{t_r E_2}\right) + B\sin\beta\left(\frac{P}{t_r E_2}\right)\tan\phi_r\right\}$$

$$+ j\left\{(A t_r E_2)\sin\alpha - B\cos\beta\left(\frac{P}{t_r E_2}\right)\tan\phi + B\sin\beta\left(\frac{P}{t_r E_2}\right)\right\} \qquad (7.8)$$

Equating only magnitudes in Eq. (7.8),

$$t_s^2 E_1^2 = \left\{ A t_r E_2 \cos\alpha + B\cos\beta\left(\frac{P}{t_r E_2}\right) + B\sin\beta\left(\frac{P}{t_r E_2}\right)\tan\phi_r\right\}^2$$

$$+ \left\{ A t_r E_2 \sin\alpha - B\cos\beta\left(\frac{P}{t_r E_2}\right)\tan\phi + B\sin\beta\left(\frac{P}{t_r E_2}\right)\right\}^2 \qquad (7.9)$$

The minimum transformers-tap ratio for co-ordination between the two transformers is achieved when

$$t_s t_r = 1 \tag{7.10}$$

Substituting $t_s = t_r^{-1}$ in Eq. (7.9) and simplifying,

$$E_1^2 E_2^2 = A^2 t_r^4 E_2^4 + B^2 P^2 \sec^2 \phi + 2ABP t_r^2 E_2^2 \cos(\alpha - \beta)$$
$$- 2ABP t_r^2 E_2^2 \sin(\alpha - \beta) \tan \phi$$

or
$$C_4 t_r^4 + C_5 t_r^2 + C_6 = 0 \tag{7.11}$$

when
$$C_4 = A^2 E_2^4$$

$$\left.\begin{array}{l} C_5 = 2ABP\, E_2^2 \cos(\alpha - \beta) - 2ABP\, E_2^2 \sin(\alpha - \beta) \cdot \tan\phi \\ C_6 = B^2 P^2 \sec^2 \phi - E_1^2 E_2^2 \end{array}\right\} \tag{7.12}$$

$$\therefore \quad t_r = \left[\frac{-C_5 \pm (C_5^2 - 4C_4 C_6)^{1/2}}{2C_4}\right]^{1/2} \quad \text{and} \quad t_s = \frac{1}{t_r} \tag{7.13}$$

Thus, the tap-range can be determined to achieve specified voltage control at the load bus having specified power factor.

Simulation

It is required to make an assessment of reactive power requirement for the voltage control of a long transmission line supplying a variable load of 0.8 p.f. (lag) at its end [A = 0.945 $\angle 1°$ p.u. and B = 0.095 $\angle 73°$ p.u.] The active power demand varies from 0.2 to 1.5 p.u. It is desired to maintain load voltage at nominal value ±5%.

TABLE 7.2 Results of solution of sample problem

P_r	$E = 0.95 E_s$			$E_r = E_s$			$E_r = 1.05 E_s$		
P_r	Q_r	t_r	t_s	Q_r	t_r	t_s	Q_r	t_r	t_s
0.2	1.015	1.044	0.957	0.546	1.018	0.981	0.024	0.994	1.005
0.3	0.983	1.039	0.962	0.513	1.013	0.986	−0.110	0.980	1.010
0.4	0.950	1.033	0.967	0.470	1.008	0.991	−0.047	0.925	1.015
0.5	0.913	1.027	0.972	0.443	1.003	0.996	−0.084	0.980	1.020
0.6	0.881	1.022	0.978	0.407	0.980	1.001	−0.122	0.975	1.025
0.7	0.844	1.016	0.983	0.369	0.992	1.007	−0.161	0.970	1.030
0.8	0.807	1.010	0.989	0.303	0.987	1.012	−0.201	0.965	1.035
0.9	0.768	1.005	0.995	0.290	0.982	1.018	−0.243	0.960	1.040
1.0	0.728	0.989	1.001	0.249	0.976	1.023	−0.286	0.955	1.052
1.1	0.687	0.993	1.007	0.206	0.971	1.029	−0.329	0.950	1.052
1.2	0.644	0.986	1.013	0.162	0.965	1.025	−0.375	0.945	1.057
1.3	0.600	0.980	1.019	0.117	0.959	1.041	−0.261	0.940	1.063
1.4	0.555	0.974	1.025	0.071	0.954	1.047	−0.468	0.934	1.069
1.5	0.508	0.968	1.032	0.024	0.948	1.054	−0.517	0.929	1.075

The results show a decrease in the reactive power supplied by the line as the load active power is increased. The negative value shows that the line is supplying leading reactive power within the range of voltage control. The tap-ranges determined for the voltage control within specified tolerances are also within the permissible range of tappings. It would, otherwise, be necessary in the system to inject VAR in the line to maintain the voltage at the required value.

The given method can be applied to predetermine the reactive power requirement of a long transmission line for voltage control and to study the feasibility of using a tap-changing transformer. In this context, it is important to note that the transformer does not improve the reactive power flow position but only redistributes it. Also, the current in the transmission line increases with the increase in the transformation ratio. The voltage drop also increases, offsetting the voltage increase to some extent. In EHV radial lines, the line and nodal reactances being high, it is possible for this voltage drop to be too large for the transformer to maintain voltage with the tap range available.

7.4 EFFECT OF ON-LOAD TAP-CHANGER TRANSFORMER ON VOLTAGE STABILITY

It has just been described how the secondary voltage of a transformer is usually maintained at or very near the nominal value by the operation of the tap changer, when the voltage of the primary transmission system drops. This is possible provided that the system is not in the state of extreme shortage of reactive power. However, if the load demand becomes excessively heavy, the secondary voltage may become unstable. The instability of the voltage is basically due to reactive power shortage and the entire mechanism of voltage collapse during this condition has already been explained earlier.

The situation further deteriorates with the adverse role played by the induction motors as also indicated earlier. Under this condition, raising the position of on-load tap-changer in an attempt to raise the secondary voltage of the load bus will not work and the bus voltage will gradually collapse. To understand this problem from a quantitative point of view, let a simple model of the power line be assumed with resistive load at the receiving-end side (Figure 7.5) and constant sending-end voltage. Here, the equivalent circuit of Figure 7.5 has been shown in Figure 7.6. The secondary voltage is then given as

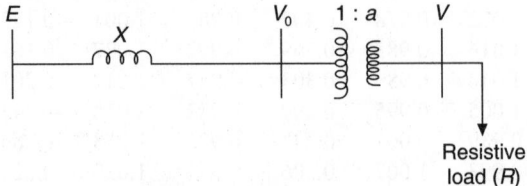

Figure 7.5 A simple model of power line with resistive load.

$$|V| = \frac{R}{\sqrt{R^2 + (a^2 X)^2}} aE \tag{7.14}$$

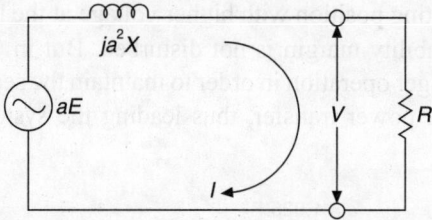

Figure 7.6 Equivalent circuit of Figure 7.4.

To find the sensitivity of the voltage with tap change, the ratio $\partial |V|/\delta a$ is to be obtained.

$$\therefore \quad \frac{\partial |V|}{\delta a} = \frac{R(R + a^2 X)(R - a^2 X)}{[R^2 + (a^2 X)^2]^{3/2}} E \tag{7.15}$$

To have stable voltage state,

$$\frac{\partial |V|}{\delta a} > 0$$

This is only possible if $\quad R > a^2 X \tag{7.16}$

The above Eq. (7.16) shows that the secondary voltage drops if the tap position 'a' is raised in order to boost the load bus voltage. At the condition when $a^2 X > R$, the voltage stability is lost. In case of EHV transmission line, X being higher in magnitude, voltage instability may spontaneously occur when an attempt is made to raise the tap position. With enhancement in a, $(a^2 X)$ factor further increases showing that even if the load bus voltage is raised, the system is pushed towards voltage instability. Voltage instability will spontaneously occur, under this condition when at any moment the magnitude of $a^2 X$ becomes equal to R.

It has been analytically established that the tap changer does not serve the purpose of maintaining voltage stability once the system operation enters into a critical state. The following discussion relates the same findings through qualitative discussion.

Figure 7.7 represents the *V-P* curve of a typical transmission system for different tap positions neglecting the effect of induction motors on the load bus, i.e. taking only the static impedance loads. It may be observed that if the initial system operation be considered at point A_0 by tap changing, the operating point first shifts to A_0'. At point A_0' the voltage is also slightly enhanced and the system is capable of transmitting more power though the system operation approaches the voltage stability limit. In case tap changing proceeds further, the point of system operation reaches A', thus further increasing the power transfer capability. Thus the tap changer operation, though enhances or maintains the secondary bus voltage and increases power transfer limit, pushes the system towards instability. In case the level of power transfer is returned to the original state, the system may settle at a new operating position with higher voltage at the load bus (point A_0''). Here the voltage stability margin is not disturbed. But in case of voltage dips (point A_1'), tap-changer operation in order to maintain the secondary bus voltage, mostly enhances the power transfer, thus leading the system towards stability limits.

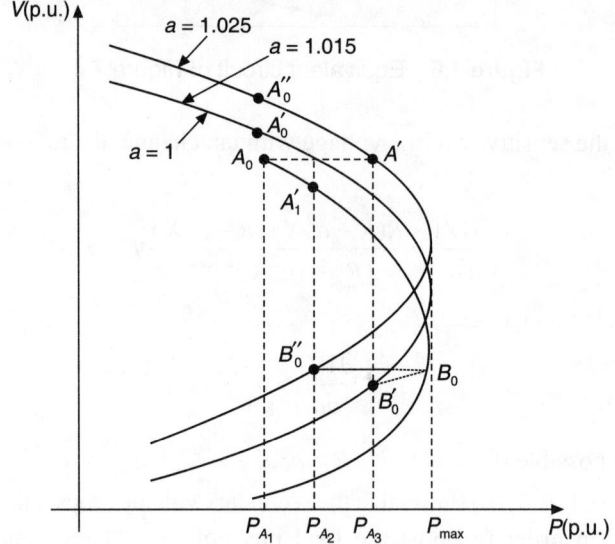

Figure 7.7 V–P characteristic with static impedance load at receiving-end bus.

It is clearly evident from Figure 7.7 that any tap-changer operation in the unstable zone of the characteristics does not elevate the bus voltage. Instead, it pushes the system more towards the unstable zone. This is evident from the shifting of point of operation from B_0 (initial operating point) to B_0'' (or B_0') in the figure. The power system also loses the capability of power transfer. Further tap change in this zone will further reduce the power transfer capability and also will lead to voltage collapse. Thus, in case the system operation lies below the stability limit (denoted by the nose of the curves) tap changer operation, in order to enhance the secondary bus voltage, leads to voltage collapse. Its operation above the stability limit may be stable even when the system is pushed toward voltage instability.

In case dynamic loading is assumed to be present in the load bus, the V-P characteristic for transformer operation is shown in Figure 7.8. The system operation is assumed to take place at A_0 where voltage stability is retained. In encountering any voltage crisis at load bus, the transformer tap changing take place thus, pushing the system towards point A'. Assuming the real power loading to be constant, the final state of operation is pointed at A_0'. Hence the system settles down at a new stable position. On the contrary, if the initial operating point lies below the voltage stability limit, tap changing results in voltage collapse. In this case, the reactive power intake of the induction motor at load bus goes on increasing and thus pulls down the state of operation to complete voltage collapse.

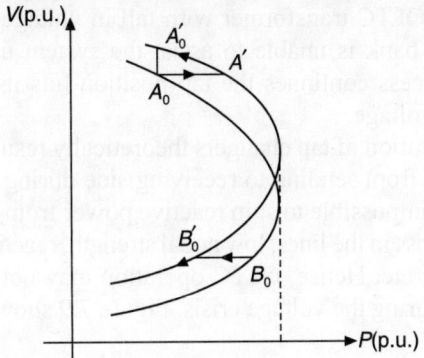

Figure 7.8 *V-P* characteristic with transformer operation with dynamic loads.

The role of OLTC transformer is important as it has been shown that variation in the tap position of OLTC transformer changes the value of r_{opt} (i.e. the ratio of magnitude of source and load reactances) thus critically affecting the voltage stability of the power system. It has also been shown that the operation of OLTC does not increase the value of limiting power angle effectively. Hence until and unless the reactive power status at the load bus improves, OLTC operation may yield reverse action during the course of enhancement of bus voltage using OLTC. Further, the critical bus voltage deteriorates with tap-changer operation. Also the OLTC operation, in order to enhance the load bus voltage, virtually enhances the magnitude of optimum value of source-to-load reactance which severely affects voltage stability.

7.5 PRACTICAL ASPECTS OF VOLTAGE INSTABILITY DUE TO OLTC OPERATION

OLTC operation can be both a boon and a curse in voltage control as has been described in the previous discussion. This operation, in a normal state can maintain the desired voltage but can also increase the risk of a voltage collapse. If the system reactive power position is not improved, the operation of OLTC may be able to restore distribution load voltage for sometime while leading the transmission system towards voltage instability.

Though theoretically voltage collapse in a weak power system takes place at around 70% of the rated voltage, practical reports suggest that voltage collapse may occur at around 80% or even above the value of rated voltage. This is due to the adverse contribution of loads to the load bus voltage. The dynamic characterisation of loads imposes premature voltage instability.

It is a common trend to use shunt capacitors in the upstream side of the receiving-end system to assist the system to recover from voltage stability problems. However, an insight into the problem reveals that this scheme may not yield the desired result to encounter voltage crisis in the load bus. In this case the operation of OLTC maintains the reactive power demand (as well as the real power demand) of the load constant. However, the capacitor output drops at the primary side of the OLTC transformer with fall in voltage on the primary side. Thus, the capacitor bank is unable to assist the system under a voltage crisis problem. If this process continues the tap position hits its limit causing sharp decrease in system voltage.

Though the operation of tap changers theoretically results in enhancement in reactive power flow from sending to receiving side during voltage problems, in practice it is almost impossible to ship reactive power from source to load due to high series reactive loss in the lines, low nodal strength, reactive power constraints, generator limitation, etc. Hence, OLTC operation may not yield any effect in a real power system during the voltage crisis. Figure 7.9 shows this condition for a heavily loaded line.

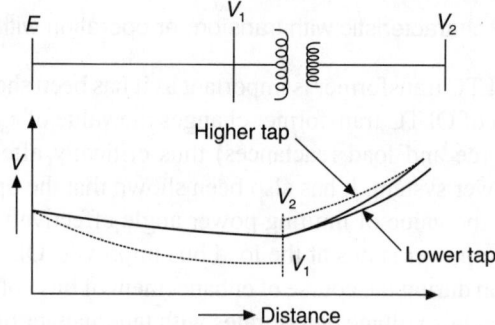

Figure 7.9 Voltage profile in a heavily loaded line.

It is now evident that the operation of OLTC transformers improves the load bus voltage provided the system operation is within the voltage stability limit. For static impedance loads, OLTC operation in this zone enhances the real power demand which again requires further tap change. Continuous tap-changer operation may lead to voltage instability as has been shown in the foregoing discussions. Operation of the tap-changer at or beyond the voltage stability limit clearly leads the system to voltage collapse. Load contribution being considered the operation of OLTC in case the system voltage is close to stability limit, leads to system voltage instability. OLTC operation does not enhance critical power angle and deteriorates critical voltage. It also adversely affects the system by increasing the ratio of source-to-load reactance. Capacitive reactive power injection on the primary side of OLTC transformers also serves little in enhancing voltage stability. OLTC operation is also least effective in transporting reactive power from source to load through long lines in order to boost the load bus voltage.

Hence most of the power utilities have found that OLTC operation in the bulk system network offers limited assistance and the utilities have relegated the

OLTC operation under manual control. OLTC transformers, except for a small range of voltage control, have been given a low priovity in the list of remedial actions to solve voltage stability problems.

7.6 METHODS OF IMPROVING VOLTAGE STABILITY

Voltage stability can be improved by adopting the following means:
  (a) Enhancing the localised reactive power support using passive shunt compensators
  (b) Compensating the line length.
  (c) Shedding the load during contingencies.
  (d) Erecting additional transmission lines.
  (e) Using FACTs controllers

Out of these five alternatives, the schemes in (a) and (b) are mostly preferred due to their efficiency and economics. Load shedding is least preferred though sometimes it is very effective in maintaining voltage stability. Erection of additional transmission lines is a viable, though costlier, alternative, to improve voltage stability. However, financial constraints mostly set limits on this option in order to implement it. In recent installations, alternative (e) is becoming a reasonable choice as it improve voltage stability and can control the real power flow too.

7.7 ENHANCEMENT OF LOCALISED REACTIVE POWER SUPPORT

Load bus being most susceptible to voltage instability, localised reactive support improves using passive devices voltage stability. Local reactive power support can be given by utilising (a) shunt capacitor banks and (b) synchronous condenser.

7.7.1 Fixed Shunt Capacitor Compensation at the Receiving-end

Receiving-end voltage profile

Analysing the Jacobian of the real and reactive power flow equations for a lossless uncompensated transmission line model, line(s) being mostly reactive power constrained, the receiving-end voltage can be expressed as

$$V = E(\cos \delta - \sin \delta \tan \phi) \qquad (7.17)$$

Taking into account the operation with fixed capacitive compensation applied at the load (receiving) end, the modified reactive power expression at the receiving-end is given by

$$Q = \frac{EV}{x_l} \cos \delta - \frac{V^2}{x_l} + V^2 \omega C \qquad (7.18)$$

which on simplification yields

$$V = \frac{E(\cos\delta - \tan\phi \sin\delta)}{(1 - x_l \cdot \omega C)} \tag{7.19}$$

and

$$\tan\phi = \frac{E \cdot \cos\delta - V(1 - x_l \cdot \omega C)}{VE \cdot \sin\delta} \tag{7.20}$$

Figures 7.10 (a), (b) and (c) represent the profile of P_{max} for variable shunt compensation and line length as well as system voltage characteristics with varying degree of shunt capacitive support at different power factors. These figures show the improvement with shunt capacitor compensation.

Continuous enhancement of the capacitance value increases the system voltage persistently. However, at a very high degree of support, the receiving-end bus experiences over-voltage. Hence, this phenomenon is of academic interest only as no practical system can sustain such a high voltage. Optimum selection of shunt capacitance is required for proper voltage control at the load bus.

Figure 7.11(a) represents the effect of shunt compensation on the ratio of the load reactive power factor and load power factor (i.e. $\tan\phi$). The nonlinearity of

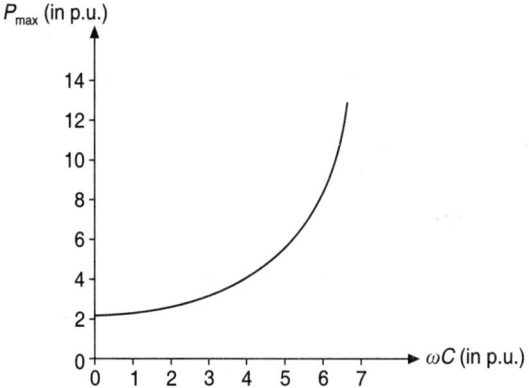

Figure 7.10(a) Profile of maximum power transfer for different shunt capacitances.

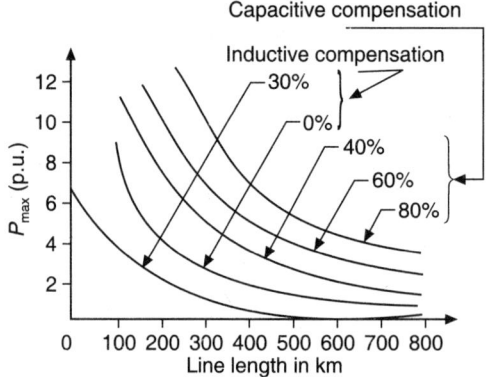

Figure 7.10(b) Maximum power transfer as affected by shunt compensations.

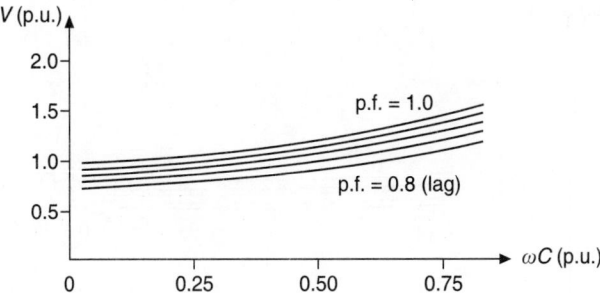

Figure 7.10(c) Receiving-end voltage improvement for terminal capacitances.

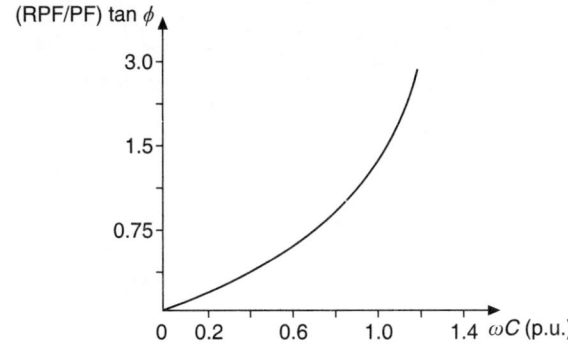

Figure 7.11(a) Effect of shunt compensation on the ratio of (RPF/PF).

the curve indicates the pronounced effect of capacitive compensation on the above mentioned ratio and indicates the improvement of load power factor with capacitive support.

[However, when the shunt reactor compensation is applied at the load end, the reactive power flow equation of a shunt reactor compensated system is given by

$$V = \frac{E(\cos\delta - \sin\delta \tan\phi)}{1 + (x_l/\omega L)} \qquad (7.21)$$

Figure 7.11(b) shows the effect of increasing the shunt reactance on the receiving-end voltage profile of a shunt compensated system. The receiving-end voltage decreases with the increase in shunt reactance indicating its applicability in the lightly-loaded system. The shunt reactive compensation thus adjusts the receiving-end voltage to the rated value compensating for the 'Ferranti Effect' for a lightly-loaded EHV system.

Voltage stability limit when shunt capacitor compensation is applied

It has been shown earlier that the determinant of the Jacobian [J] of the load flow equations of any uncompensated line model is given by

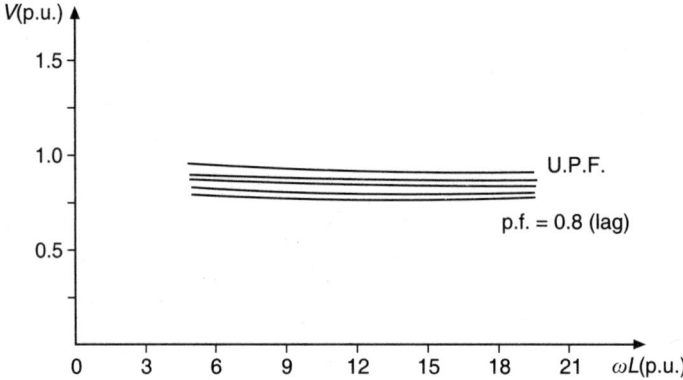

Figure 7.11(b) System voltage characteristics at different power factors for varying degree of reactive compensation.

$$\Delta[J] = \frac{1}{x_l^2}(-E^2 V + 2V^2 E \cos \delta) \quad (7.22)$$

But $[J]$ will be singular if $\Delta[J]$ be equated to zero signifying single solution of the load flow equations. This condition determines the voltage stability limit, and simplification of Eq. (7.22) with $\Delta[J] = 0$, at voltage stability limit yields

$$V_{cri} = \frac{E}{2 \cos \delta} \quad (7.23)$$

Again, the Jacobian $[J]$ of the load flow equation of a shunt capacitor compensated line model can be analysed in the same way, to observe the effect of shunt capacitor compensation on the critical voltage when the receiving-end bus voltage of stability limit is given by

$$V_{cri} = \frac{E}{2 \cos \delta \, (1 - x_l \cdot \omega C)} \quad (7.24)$$

Figure 7.12 represents the profile of the critical voltage for varying degrees of power angle at a constant source voltage.

Applying shunt capacitive compensation, the profile for the critical value of the voltage stability limit is obtained in Figure 7.13 against the magnitudes of power angle at varying degrees of compensation. Voltage support at voltage stability limit is better in a shunt capacitor compensated system if the terminal capacitance is properly selected.

On the other hand, when shunt reactor compensation is applied at the load bus, analysing the Jacobian of the load flow equations in the same manner, the critical voltage at the voltage stability limit is given by

$$V_{cri} = \frac{E}{2 \cos \delta \, (1 - x/x_{SH})} \quad (7.25)$$

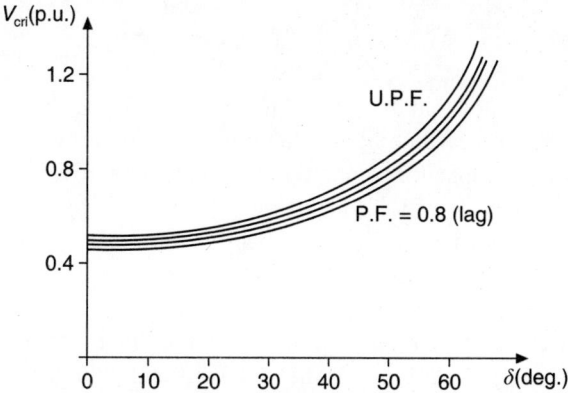

Figure 7.12 Profile of critical receiving-end voltage at different power factors for varying power angle.

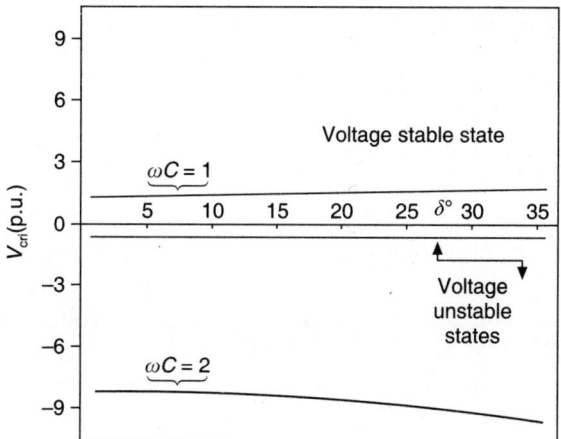

Figure 7.13 Profile showing the voltage stable and unstable states with capacitive compensations.

Equation (7.25) reveals that the critical voltage at the receiving-end deteriorates with increase in shunt reactor compensation.

Critical power angle at voltage stability limit

Taking into account the uncompensated line model, substituting the value of V_{cri} from Eq. (7.17) into Eq. (7.23),

$$\delta_{cri} = \frac{\pi}{4} - \frac{\phi}{2} \qquad (7.26)$$

In a shunt capacitor compensated line model, δ_{cri} can be obtained in a similar way by substituting the value of V_{cri} from Eq. (7.24) into Eq. (7.17) and is given by

$$\delta_{cri} = \frac{\pi}{4} - \frac{\phi_{new}}{2} \qquad (7.27)$$

It has been earlier stated that though the magnitude of and δ_{cri} in Eq. (7.26) seems to be the same, still the value of δ_{cri}, the power angle at voltage stability limit, improves with shunt capacitor compensation due to an improvement of load power factor (designated by ϕ_{new}) with inclusion of the shunt capacitor. On the other hand, the critical power angle decreases with shunt reactor compensation since the inclusion of the shunt reactor at load end deteriorates the load power factor. Hence the inclusion of shunt reactor has got no practical interest, except for controlling the receiving-end bus voltage when the line is lightly loaded.

The above analysis reveals that the receiving-end voltage profile and power factor improve utilising the shunt capacitor support at the receiving-end while they deteriorate with inductive support at the load end. Voltage stability limit of a shunt capacitive compensated power system is enhanced remarkably in comparison to an uncompensated or reactor compensated system. However, proper selection of the shunt capacitance is important in restoring the voltage stability. With excessive high values of shunt capacitance, the system may encounter an over-voltage state. Shunt capacitive compensation, on improving the load end power factor, also increases the magnitude of the critical power angle at voltage stability limit. Hence shunt capacitor compensation is a viable means to improve both the voltage profile and the voltage stability limit, in a reactive power-constrained transmission system.

Referring to the discussion of stability margin, it may be observed that though shunt capacitor compensation enhances the receiving-end voltage profile, power angle and critical voltage at the receiving-end bus, the voltage stability margin deteriorates. Moreover, with enhancement of shunt capacitance at the load bus, its capability to meet more and more real power demand is enhanced, and hence the common trend becomes enhancement of the receiving-end bus loading under the influence of enhanced shunt capacitive support. However, it may be observed that though the voltage profile can be kept near the nominal value by enhancing the shunt capacitive support (vide Figure 7.14(a) where it has been shown that the receiving-end voltage remains at near rated value with switched shunt capacitor compensation and with enhancement of power flow), the system operation may enter the unstable state leading to steady state instability (Figure 7.14(b)). The only viable alternative is then to install a new line.

7.8 SERIES COMPENSATION

Series compensation is provided by installation of capacitor banks in each phase and in series with the power trasmission line. It not only reduces the line reactance but also increase the power transfer as well as the steady state stability limit. It virtually increases the surge impedance loading. In long distance power transmission, series capacitor compensation is beneficial provided proper care is

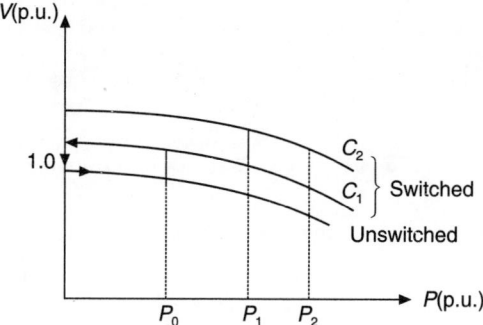

Figure 7.14(a) Receiving-end bus voltage with shunt capacitive support.

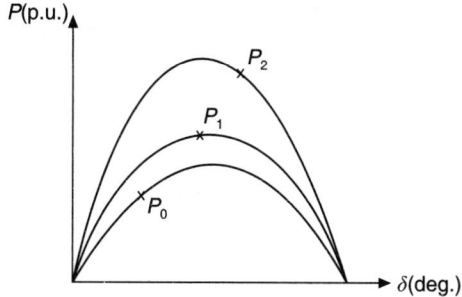

Figure 7.14(b) Steady state instability even with shunt compensation.

taken to mitigate the subsynchronous resonance problem. Usually the degree of series compensation (X_C / X_L) is 40–80% as over-compensation may give rise to steady state over-voltage and low damping. In order to model the series compensation, we assume a lossless symmetrical line.

7.8.1 Concepts in Loadability and Voltage Stability of Series Compensated Lines

Criterion for line loadability limit

The active and reactive power flows at the receiving-end bus for a transmission line equipped with series compensation and OLTC transformer, shown in Figure 7.15, are given by:

$$P_R = \frac{E \cdot V \cos(\beta_0 - \delta)}{a |B_0|} - \frac{A_0 V^2 \cos(\beta_0 - \alpha_0)}{a^2 |B_0|} \quad (7.28)$$

$$Q_R = \frac{E \cdot V \sin(\beta_0 - \delta)}{a |B_0|} - \frac{A_0 V^2 \sin(\beta_0 - \alpha_0)}{a^2 |B_0|} \quad (7.29)$$

The sending-end and receiving-end transformer reactances have also been included in $|A_0|$ and $|B_0|$. The 'a' is the status of the receiving-end transformer tap ratio.

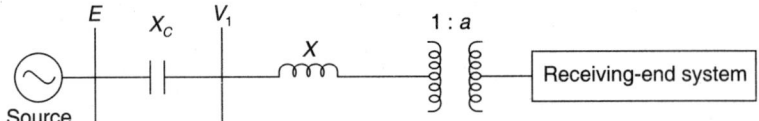

Figure 7.15 A series capacitor componsated line.

P_R and Q_R can also be obtained in terms of voltage dependent load characteristics, i.e.

$$P_R = P_R^0 \cdot V^m \tag{7.30}$$

$$Q_R = Q_R^0 \cdot V^n = P_R^0 \tan\phi V^n \tag{7.31}$$

where P_R^0 and Q_R^0 are the real and reactive loadings of the line corresponding to $V_R = 1.0$ p.u., ϕ is the power factor angle and m and n are the characteristic parameters of the load. For a given value of the degree of series compensation, the maximum power transfer criteria, i.e. the criterion of the line loadability limit considering the voltage dependence of loads is determined by considering the singularity of the Jacobian J of Eqs. (7.28) and (7.29) in conjunction with Eqs. (7.30) and (7.31) and is given by

$$\frac{\partial f_1}{\partial \delta} \cdot \frac{\partial f_2}{\partial V} - \frac{\partial f_1}{\partial V} \cdot \frac{\partial f_2}{\partial \delta} = 0$$

or

$$E - \cos(\beta_0 - \delta) \cdot \frac{a|B_0|mP_R}{V} + \frac{2|A_0|V \cos(\delta_0 - \alpha_0)}{a}$$

$$- \sin(\beta_0 - \delta) \cdot \frac{a|B_0|nQ_R}{V} + \frac{2|A_0|V \sin(\delta_0 - \alpha_0)}{a} = 0 \tag{7.32}$$

Reactive power requirement at the load-end when the angular separation reaches the limiting value

The reactive power requirement at the receiving bus, to allow an angular separation between the two ends of the system of Figure 7.16 with the load-end voltage $E = V$, can be obtained analytically; the critical power transfer $P_{R\text{crit}}$ is obtained from Eq. (7.28) by substituting $E = V$ and $\delta = \beta_0$ and is given by

$$P_{R\text{crit}} = \frac{E^2}{a|B_0|}\left[1 - \frac{|A_0|\cos(\beta_0 - \alpha)}{|a|}\right] \tag{7.33}$$

The corresponding critical reactive power $Q_{R\text{crit}}$ can similarly be obtained from Eq. (7.29) and is given by

$$Q_{R\text{crit}} = \frac{|A|E^2 \sin(\alpha - \beta)}{a^2|B_0|} + Q_{\max} \tag{7.34}$$

where $Q_{\max}$ is the required var reserve at the load-end when the transmission angle reaches its limiting value β_0. Further, $P_{R\text{crit}}$ and $Q_{R\text{crit}}$ are also obtained

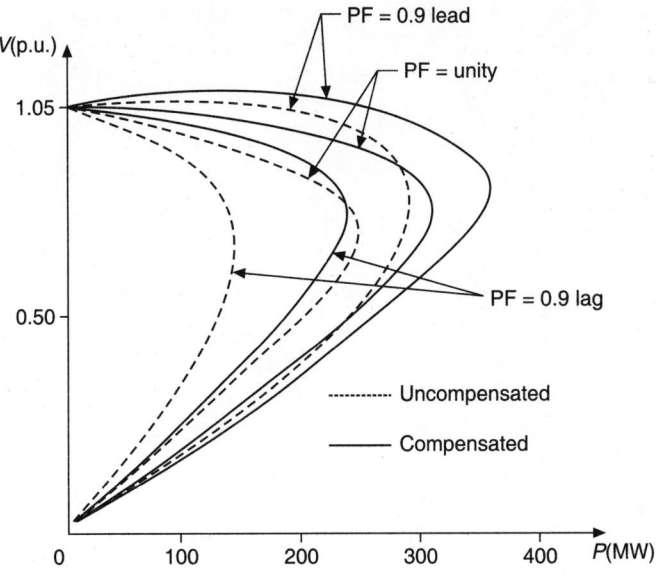

Figure 7.16 Receiving-end voltage characteristics of a 220 kV 300 kms line, $V_s = 1.05$ p.u.

from the voltage dependent load characteristics, given by Eqs. (7.30) and (7.31), by substituting $E = V$ and replacing P_R^0 and Q_R^0 by $P_{R_{crit}}^0$ and $Q_{R_{crit}}^0$ respectively. Thus from Eqs. (7.30) and (7.31), simplification yields

$$Q_{R_{crit}} = P_{R_{crit}} \tan\phi \, E^{n-m} \tag{7.35}$$

Substituting the values of $P_{R_{crit}}$ and $Q_{R_{crit}}$ from Eqs. (7.29) and (7.30) into Eq. (7.31),

$$Q_{R_{crit}} = \frac{E^2}{a|B_0|}\left[1 - \frac{|A_0|\cos(\beta_0 - \alpha_0)}{a}\right]\tan\phi\, E^{n-m} + \frac{A_0 E^2 \sin(\beta_0 - \alpha_0)}{|B_0|a^2} \tag{7.36}$$

If the available VAR reserve at the load-end is less than Q_{max}, the critical angular separation δ_{crit} will be less than the limiting value β_0. In such a case, the value of δ_{crit} is dependent on whether the load bus is voltage controlled or not. The relevant expressions for δ_{crit} with VAR reserve less than Q_{max} for the cases of voltage controlled or voltage uncontrolled buses are derived in the next section.

Critical angular separation δ_{cri} with finite VAR reserve

(i) *Voltage controlled load bus* The condition is to keep V equal to E for a given degree of series compensation K with a VAR reserve 'min Q_{max}' supplied by either a synchronous condenser or a static VAR source at the load bus. If the available 'min Q_{max}' at the receiving-end bus is not large enough to allow δ to reach its limiting value β_0, the angle δ_{max} corresponding to 'min Q_{max}' is determined from the following equation:

$$\frac{E^2}{a|B_0|}\left[\cos(\beta_0-\delta_{max}) - \frac{|A_0|\cos(\beta_0-\alpha_0)}{a}\right]\tan\phi \, E^{n-m}$$

$$= \frac{E^2}{a|B_0|}\left[\sin(\beta_0-\delta_{max}) - \frac{|A_0|\sin(\beta_0-\alpha_0)}{a}\right] + \min Q_{max} \quad (7.37)$$

(ii) *Voltage uncontrolled load bus* If the line loadability limit is to be increased further from that corresponding to an angular separation of δ_{max} for a given limiting value 'min Q_{max},' the line may or may not be able to deliver the additional power. Since the bus receiving power is an uncontrolled voltage bus, the critical condition is likely to be decided by the instability of voltage at the load end rather than by the angular separation at the two ends of the system. The criterion which governs the critical operating condition under the given conditions of a limiting value of 'min Q_{max}' and a voltage uncontrolled load bus is obtained by substituting the values P_R and Q_R given in Eqs. (7.28) and (7.29) respectively in Eq. (7.32). V and δ represent the values of the corresponding parameters under the critical conditions for the transmission system with limited VAR reserve 'min Q_{max}' and an uncontrolled voltage bus at the load end. The criterion is given by

$$E - K_1 \cos(\beta_0 - \delta) - K_2 \sin(\beta_0 - \delta) = 0 \quad (7.38)$$

where

$$K_1 = mE\cos(\beta_0 - \delta) + \frac{A_0 V \cos(\beta_0 - \alpha_0)}{a}(2-m) \quad (7.39)$$

$$K_2 = nE\sin(\beta_0 - \delta) + \frac{|A_0|V\sin(\beta_0 - \alpha_0)}{a}(2-n) \quad (7.40)$$

Furthermore, using an equation similar to Eq. (7.36), the reactive power 'min Q_{max}' required at the load end for an angular separation δ is obtained from

$$\left[\frac{EV\cos(\beta_0-\delta)}{a|B_0|} - \frac{A_0 V^2 \cos(\beta_0-\alpha_0)}{a^2|B_0|}\right]\tan\phi \, V^{n-m}$$

$$-\left[\frac{EV\sin(\beta_0-\delta)}{a|B_0|} - \frac{A_0 V^2 \sin(\beta_0-\alpha_0)}{a^2|B_0|} + \min Q_{max}\right] = 0 \quad (7.41)$$

The load-end voltage V and the angular separation δ can be obtained by solving Eqs. (7.34) and (7.41) respectively.

For a given set of E, K and the line parameters, the critical angular separation δ_{cri} corresponding to the line loadability limit is given by δ_{max} or δ. Thus,

$$\delta_{cri} = \begin{cases} \beta_0, \text{ if min } Q_{max} > Q_{max} \\ \max(\delta_{max}, \delta), \text{ otherwise} \end{cases} \quad (7.42)$$

If $\delta_{cri} = \beta_0$, the line loadability is governed by angular stability criterion, otherwise, it is governed by the criterion of voltage stability.

For a given operating point, the stability margin $(S \cdot m_0)$ can be obtained as follows:

$$S \cdot m_0 = \left(1 - \frac{P_{R0}}{P_{Rcri}}\right)$$

where P_{R0} is the initial operating point power and P_{Rcri} is the line loadability limit under steady state condition.

Simulation has been carried out for a typical series compensated power system with the concept of perfect voltage control at the sending end side ($E = 1$ p.u.). The loadability limit of the transmission system with the given data, and the VAR reserve required to maintain an acceptable voltage leve of 0.95 p.u. and 30% stability margin are determined. Both voltage-dependent and voltage-independent characteristics of the load are considered. For the voltage-dependent loads, the load characteristic parameters considered are $mn = 0.5$ and three load power factors, namely, 0.86 lead, unity P.F. and 0.82 lag have been considered. Tables 7.3 and 7.4 give the computational results lor voltage-independent constant power factor loads and Tables 7.5 and 7.6 give the computational results for voltage-dependent loads utilising the theory presented.

Table 7.3 shows the computational results corresponding to a loading of 0.5 times SIL. The effects of series compensation and load power factor on the line loadability limit can be seen from this table. δ_{cri}, δ_M and δ_0 represent the angular separations for the system operating respectively with zero, 30% and greater than 30% margins of stability. It can be observed from Table 7.3 that δ_M is independent of the load power factor but δ_0 and δ_{cri} are dependent on it. For leading power factor loads, higher stability margins are achieved compared to unity and lagging power factor loads. The stability margin also increases with the degree of series compensation. The critical load-end voltage V_{cri} is higher for leading power factor loads than for lagging and unity power factor loads. For leading power factor and unity power factor loads, the minimum VAR reserve 'min Q_{max}' required at the load-end to maintain an acceptable voltage level and a desired stability margin is negative, indicating the necessity for a reactive power sink at the receiving bus. Since 'min Q_{max}' requirement varies according to the load power factor and the degree of series compensation, a static VAR system is desirable.

Table 7.4 shows the computational results for a line loading of 1.5 SIL. It can be observed that the uncompensated line is angle limited for a loading of 1.5 SIL and assumed load power factors, because δ_{cri} reaches its limiting value β_0. The same line becomes voltage limited when the degree of series compensation is 50% since δ_{cri} is now less than β_0. In the latter case, the voltage would collapse below the critical value before δ_{cri} could reach the limiting value β_0.

Tables 7.5 and 7.6 show the results of studies considering voltage-dependent load characteristics. Table 7.5 corresponds to the unity ratio for the transformer tap. Table 7.5 gives the results corresponding to three different transformer tap

Table 7.3 Effect of load power factor on line loadability limit and VAR source allocation $P_{R_0} = 0.5$ SIL, $P_{R\max} = 0.714$ SIL

S. No.	PF	K_S	δ_0 (deg)	V_0 (pu)	Q_0 (pu)	$\delta_{\max}$ (deg)	$V_{\max}$ (pu)	δ_{cri} (deg)	V_{cri} (pu)	P_{cri} (pu)	$s.m_0$	$\min_{Q\max}$ (pu)
1	1.0	0.0	15.926	1.00	−0.117	24.330	0.95	39.310	0.755	0.550	0.450	0.099
2		0.5	11.400	0.984	−0.144	16.970	0.95	38.400	0.710	0.712	0.560	−0.144
1	0.86 lead	0.0	17.460	0.910	−0.370	24.330	0.95	46.690	0.850	0.704	0.560	−0.370
2	0.82 lag	0.5	12.170	0.911	−0.412	16.970	0.95	50.030	0.863	1.053	0.703	−0.412
1		0.0	15.926	1.00	0.102	24.300	0.95	36.790	0.730	0.500	0.376	0.214
2		0.5	11.250	1.00	0.095	16.970	0.95	32.910	0.662	0.586	0.466	0.168

Table 7.4 Effect of power factor on line loadability limit and VAR source allocation $P_{R_0} = 1.5$ SIL, $P_{R\max} = 2.143$ SIL

S. No.	PF	K_S	δ_0 (deg)	V_0 (pu)	Q_0 (pu)	$\delta_{\max}$ (deg)	$V_{\max}$ (pu)	δ_{cri} (deg)	V_{cri} (pu)	P_{cri} (pu)	$s.m_0$	$\min_{Q\max}$ (pu)	β_{0I} (deg)
1	1.0	0.0	56.180	1.00	0.360	88.620	1.00	88.620	1.00	1.120	0.16	0.980	88.62
2		0.5	36.110	1.00	0.180	58.090	1.00	58.090	1.00	1.340	0.30	0.640	87.82
1	0.86 lead	0.0	56.180	1.00	−0.200	88.620	1.00	88.620	1.00	1.120	0.16	0.980	88.62
2	0.82 lag	0.5	36.110	1.00	−0.390	58.090	1.00	58.090	1.00	1.340	0.30	−0.162	87.82
1		0.0	56.180	1.00	0.980	88.620	1.00	88.620	1.00	1.120	0.16	0.980	88.62
2		0.5	36.110	1.00	0.83	58.090	1.00	58.090	1.00	1.340	0.30	1.580	87.82

Table 7.5 Effect of load characteristics on line loadability limit and VAR source allocation $m = n = 0.5$ $PF = 0.82$ lag

S. No.	P_{R0} (SIL)	K_S	δ_0 (deg)	V_0 (pu)	Q_0 (pu)	δ_{max} (deg)	V_{max} (pu)	δ_{cri} (deg)	V_{cri} (pu)	P_{cri} (pu)	$s.m_0$	$\min Q_{max}$ (pu)
1	0.5	0.0	15.93	1.00	0.100	24.33	0.95	46.77	0.57	0.47	0.34	0.21
2		0.5	11.25	1.00	0.095	16.97	0.95	42.28	0.50	0.55	0.43	0.17
1	1.0	0.0	33.42	1.00	0.470	56.26	0.95	63.46	0.87	0.88	0.29	0.92
2		0.5	23.04	1.00	0.420	36.09	0.95	56.78	0.67	0.90	0.31	0.72

Table 7.6 Effect of receiving-end transformer tap on line loadability limit and VAR-source allocation with voltage-dependent loads $P_{R0} = $ SIL $m = n = 0.5$ $PF = 0.82$ lag

S. No.	N	K_S	δ_0 (deg)	V_0 (pu)	Q_0 (pu)	δ_{max} (deg)	V_{max} (pu)	δ_{cri} (deg)	V_{cri} (pu)	P_{cri} (pu)	$s.m_0$	$\min Q_{max}$ (pu)
1	0.95	0.0	31.214	1.00	0.454	49.678	0.95	64.647	0.863	0.899	0.311	1.012
2		0.5	21.934	1.00	0.596	33.961	0.95	61.720	0.653	1.028	0.395	0.995
1	1.00	0.0	33.420	1.00	0.470	56.260	0.95	63.460	0.870	0.876	0.290	0.920
2		0.5	23.040	1.00	0.420	36.090	0.95	56.780	0.670	0.900	0.310	0.72
1	1.05	0.0	34.690	1.00	0.517	61.548	0.95	62.285	0.906	0.829	0.246	0.889
2		0.5	23.595	1.00	0.344	37.173	0.95	53.884	0.700	0.839	0.259	0.573

Subscript '0' indicates the initial operating condition, 'max' indicates the system parameters operating with 30% stability margin and 'cri' represent the same with zero stability margin.

ratios. By comparing Tables 7.5 and 7.3, it can be observed that the line loadability limit and the minimum VAR reserve required to maintain an acceptable voltage level and a desired stability margin are altogether different for voltage dependent loads and for constant power factor loads. From Table 7.6, it can be concluded that a decrease in transformer tap ratio from unity improves the margin of stability and vice versa. Similarly a decrease in tap ratio from unity increases the minimum VAR resolve requirement and vice versa.

7.8.2 Effect of Series Compensation on the Voltage Instability of EHV Long Lines

The receiving-end voltage of a series compensated transmission line has been shown to be a polynomial equation and is given by

$$V^4 + BV^2 + C = 0 \tag{7.43}$$

where

$$B = \frac{2[Q_R(a_1 b_2 - a_2 b_1) + P_R(a_1 b_1 + a_2 b_2) - E^2/2]}{a_1^2 + a_2^2}$$

and

$$C = \frac{(b_1^2 + b_2^2)(P_R^2 + Q_R^2)}{a_1^2 + a_2^2}$$

Figure 7.17 shows the variation of the receiving-end voltage with the receiving-end power for a 300 km 220 kV series compensated transmission line. For the compensated case, a compensation of 40% is considered and the series capacitor bank is located at the receiving-end of the line. It can be observed that the series compensation increases the steady state limit of power transfer for a given sending-end voltage and at given power factor of the load, without undue voltage depressions at the load end.

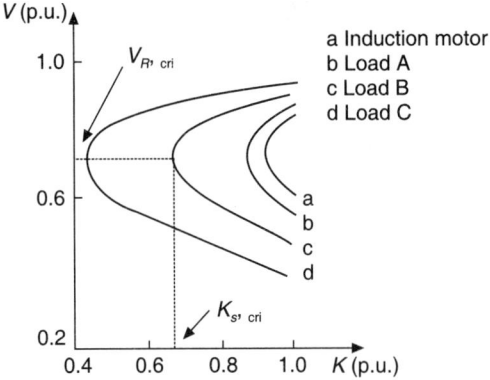

Figure 7.17 Variation of load voltage with degree of compensation.

The presented system has been assumed to have a synchronous machine with fast-acting excitation regulator connected to a 220 kV, 300 km transmission line with a receiving-end transformer (similar to that shown in Figure 7.16). The transmission line feeds the receiving-end transformer (220/132 kV on load tap-changing type) and the 132 kV bus is connected to a 33 kV bus through a 132/33 kV transformer. The characteristics of the loads are evaluated at the 132 kV side. The load has been assumed to change from 0.05 p.u. to 0.25 p.u. in 30 min while the tap control of the transformer at the receiving-end acts instantaneously. The effect of the steep increase in the load demand is to cause voltage depression at the load end. This may invariably result in voltage instability if the load demand is uncontrolled. The voltage instability occurs at an instant when the receiving-end voltage becomes more sensitive to the load demand.

The investigation revealed that there is less likelihood of voltage instability occurring in series-compensated EHV lines. The voltage dips at the receiving-end of the line depend on the transformer tap control at the load end, the type of load, the degree of series compensation and the switching time of series capacitors at the load end. The performance of the line with an initially fixed series compensation and subsequent switching of series capacitors is superior to that of an initially uncompensated line and subsequent switching of series capacitors. It has been revealed that the probabilistic nature of the load parameters and the loading conditions affect the performance of uncompensated lines more than that of compensated lines. It can be concluded from this study that the performance of EHV lines with series compensation is superior to that of an uncompensated line from the point of view of voltage stability.

7.8.3 Relation between the Receiving-end Voltage and the Degree of Compensation

It has been established in the research that the voltage stability at the load end of a series compensated line for a given mixture of voltage sensitive loads can be maintained if the degree of compensation is such that (dV/dK) is infinite or (dK/dV) is zero, for a fixed sending-end voltage. The relationship between the receiving-end voltage and the degree of compensation of an EHV line with series compensation has been obtained for a given composition of loads by formulating a polynomial of V with coefficients which are functions of the degree of compensation, the coefficients of the load polynomials, and the transmission system parameters. For a fourth degree polynomial in V of the composite load, the resulting relationship is an eight-degree polynomial in V and is given by

$$V^8 + l_1 V^7 + l_2 V^6 + l_3 V^5 + l_4 V^4 + l_5 V^3 + l_6 V^2 + l_7 V + l_8 = 0 \quad (7.44)$$

The coefficients l_k, $k = 1, ..., 8$, except l_1, l_7 and l_8 are functions of the degree of compensation and are given by

$$l_1 = \frac{k_7}{k_8}$$

$$l_2 = \{k_6 k_9 + 2(a_5 k_{10} + a_{10} k_{11})\}/k_8$$
$$l_3 = \{k_5 k_9 + 2(a_4 k_{10} + a_9 k_{11})\}/k_8$$
$$l_4 = \{k_{12} + k_4 k_9 + 2(a_3 k_{10} + a_8 k_{11})\}/k_8$$
$$l_5 = \{k_3 k_9 + 2(a_2 k_{10} + a_7 k_{11})\}/k_8$$
$$l_6 = \{k_2 k_9 + 2(a_1 k_{10} + a_6 k_{11}) - 1\}/k_8$$
$$l_7 = k_1/k_8$$
$$l_8 = k_0/k_8$$
$$k_0 = a_1^2 + a_6^2$$
$$k_1 = 2(a_1 a_2 + a_6 a_7)$$
$$k_2 = a_2^2 + a_7^2 + 2(a_1 a_3 + a_6 a_8)$$
$$k_3 = 2(a_1^2 a_4 + a_2 a_3 + a_7 a_8 + a_6 a_9)$$
$$k_4 = a_3^2 + a_8^2 + 2(a_3 a_5 + a_8 a_{10})$$
$$k_5 = 2(a_2 a_5 + a_3 a_4 + a_7 a_{10} + a_8 a_9)$$
$$k_6 = a_4^2 + a_9^2 + 2(a_3 a_5 + a_8 a_{10})$$
$$k_7 = 2(a_4 a_5 + a_9 a_{10})$$
$$k_8 = a_5^2 + a_{10}^2$$
$$k_9 = B_1^2 + B_2^2$$
$$k_{10} = 2(A_1 B_1 + A_2 B_2)$$
$$k_{11} = 2(A_1 B_2 - A_2 B_1)$$
$$k_{12} = A_1^2 + A_2^2$$

7.8.4 Relation between *E* and *V* for Voltage Stability

Voltage stability at the load end for a given degree of compensation and a given type of load is maintained if the sending-end voltage is such that (dE/dV) is greater than 0. The real and reactive powers, P_R and Q_R at the receiving-end of an EHV line with a given degree of compensation may be written as functions of the circuit variables (E, V) and the phase angle difference (δ) between E and V. Thus

and
$$\left. \begin{array}{l} P_R = f_1(E, V, \delta) \\ Q_R = f_2(E, V, \delta) \end{array} \right\} \quad (7.45)$$

P_R and Q_K must also satisfy the load characteristic given by Eq. (7.46)

$$P_R = f_3(V) \\ Q_R = f_4(V)$$ (7.46)

for a given composition of loads. Eliminating P_R and Q_R from Eqs. (7.45) and (7.46) and considering small variations in the circuit variables and applying the (dE/dV) criterion for voltage stability, it has been revealed that the voltage stability of a series compensated line is maintained if

$$E - 2V(A_1 \cos \delta + A_2 \sin \delta) - \frac{\delta P_R}{\delta V}(B_2 \sin \delta + B_1 \cos \delta)$$

$$- \frac{\delta Q_R}{\delta V}(B_2 \cos \delta - B_1 \sin \delta) > 0 \qquad (7.47)$$

where A_1, B_1 and A_2, B_2 are the real and imaginary parts of the generalised line constants A_0 and R_0 respectively.

Graphs depicting the variation of V with K for a 275 kV, 300 km long line with various composition of loads are shown in Figure 7.18. An inadequate reactive power to meet a given load demand results in lowering of the voltage at the load end, thus increasing the line drop. The voltage at which the above process continues, resulting in voltage collapse, is known as the critical load voltage. The critical load voltage V_{cri} and the corresponding critical degree of compensation K_{cri} may be obtained from Figure 7.17 by making use of the (dV/dK) criterion.

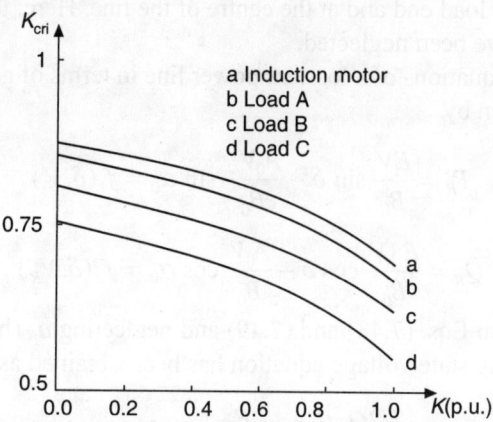

Figure 7.18 Profile of critical receiving-end voltage with p.u. series compensation for different loads.

It can be seen from Figure 7.18 thal the critical load voltage and the critical compensation have low values when the induction motor content in a load composition in reduced. It can also be seen that the critical load voltage in insensitive to P_R and Q_R and depends only upon the line and load characteristics, namely, the line constants A_0, B_0 which include the degree of compensation K, and the load coefficients $a_1, ..., a_{10}$. The critical load voltages are obtained for loads of different composition and for various degrees of compensation by the (dE/dV) criterion as well and are shown in Figure 7.18. The result obtained by (dV/dK) criterion, and the (dE/dV) criterion are in close agreement. Further, it can be observed that there is smaller possibility of voltage instability in series compensated EHV lines compared to uncompensated lines.

The impact of the variation in the sending-end voltage on voltage stability has been considered. It has been seen that decreasing the sending-end voltage to 0.9 p.u., the critical load voltage and the critical degree of compensation increases, impairing the voltage stability of the system, whereas increasing the sending-end voltage to 1.1 p.u., improves the voltage stability.

7.8.5 Effect of Location of Series Capacitor on Voltage Stability

In the following section, voltage stability of series capacitor compensated EHV lines has been presented to observe the effect of placement of series capacitor and the effect of capacitor bussing configuration on voltage stability.

The system voltage characteristic of a basic power transmission system being investigated, the expression for the critical receiving-end voltage at voltage stability limit has been presented in terms of system power angle and generalised circuit constants, utilising the criterion of singularity of the Jacobian of the load flow equations. The utility of the series capacitor compensation has also been explained and the theoretical derivations have been formulated for a 220 kV 275 km power system. Results have also been justified from the view point of application of the series capacitor at load end and at the centre of the line. Here, the effects of load characteristics have been neglected.

Power flow equations of a lossless power line in terms of generalised circuit constants are given by,

$$P_R = \frac{EV}{B_0} \sin \delta - \frac{A_0 V^2}{B_0} \sin \alpha_0 = f_1(\delta, V) \tag{7.48}$$

$$Q_R = \frac{E \cdot V}{B_0} \cos \delta - \frac{A_0 V}{B_0} \cos \alpha_0 = f_2(\delta, V_R) \tag{7.49}$$

Eliminating δ from Eqs. (7.48) and (7.49) and neglecting a_0 (being very small), the system's steady state voltage equation has been obtained as

$$V^4 + V^2 \left(\frac{2Q_R B_0 A_0 - E^2}{A_0^2} \right) + B_0^2 P_R^2 \sec^2 \phi = 0 \tag{7.50}$$

which represents a quadratic equation in V^2. The solution of Eq. (7.50) results in the expression for V given by

$$V = \left[\frac{(-2Q_R B_0 A_0 + E^2)/(A_0)^2 + \sqrt{(1/A_0^2)(2Q_R B_0 A_0 - E^2)]^2 - 4B_0^2 P_R^2 \sec^2 \phi}}{2} \right]^{1/2} \tag{7.51}$$

neglecting the negative solution, as the negative value of V does not carry any physical meaning.

Further, in a reactive power constrained line,

$$Q_R = P_R \tan \phi \tag{7.52}$$

Substituting for Q_R from Eq. (7.52) in Eq. (7.51) yields

$$V = \left[\frac{1}{2} \left\{ (-2P_R \tan \phi \, B_0 A_0 + E^2)/(A_0)^2 \right\} \right.$$

$$\left. \pm \frac{1}{2} \sqrt{\left\{ \frac{+2P_R \tan \phi \, B_0 A_0 - E^2}{A_0^2} \right\}^2 - 4B_0^2 P_R^2 \sec^2 \phi} \right] \tag{7.53}$$

Equation (7.50) represents the system's steady state voltage characteristic which, for unity power factor loading, can be reduced to

$$V = \left[\frac{E^2}{2A_0^2} \pm \frac{1}{2} \sqrt{\frac{E^4}{A_0^4} - 4B_0 P_R^2} \right]^{1/2} \tag{7.54}$$

Equation (7.53) or (7.54) indicates that the receiving-end voltage V has two real roots, one being of the higher magnitude than the other. The higher one provides the acceptable voltage profile at load end while the lower one renders the system operation at a subnormal and unacceptable voltage profile at heavy loading, causing enormous line drop and series reactive loss compounding to the lower reactive charging at this low system voltage.

The roots are real and equal when the expression under the radical sign is zero. Thus from Eq. (7.53),

$$P_R = \frac{E^2}{2B_0 A_0^2 \, (\sec \phi + \tan \phi / A_0)} \tag{7.55}$$

which at unity power factor becomes

$$P_R \text{ at upf} = \frac{E^2}{2B_0 A_0^2} \tag{7.56}$$

Equation (7.55) or (7.56) represents the critical value of real power flow to the load end when the roots of Eq. (7.53) or Eq. (7.54) are real and equal.

Considering this condition, the expression for the real and equal roots of the steady state voltage equation is given by

$$V = \left[\frac{-2P_R \tan \phi \, B_0 A_0 + E^2}{2A_0^2} \right]^{1/2} \tag{7.57}$$

or at unity power factor,

$$V = [E/(\sqrt{2} \, A_0)] \tag{7.58}$$

For a constant sending-end voltage, the Jacobian corresponding to the load flow equations is then given by

$$[J] = \begin{bmatrix} \left(-\dfrac{EV}{B_0}\cos\delta\right) & \left(-\dfrac{E}{B_0}\sin\delta + \dfrac{2A_0V}{B_0}\sin\alpha_0\right) \\ \left(-\dfrac{EV}{B_0}\sin\delta\right) & \left(-\dfrac{E}{B_0}\cos\delta + \dfrac{2A_0V}{B_0}\cos\alpha_0\right) \end{bmatrix} \quad (7.59)$$

As the determinant of this Jacobian matrix, when equated to zero, becomes singular, it signifies a single solution of the load flow equations and represents the voltage stability limit. Hence to have a voltage stable state, Del $|J| > 0$:

i.e. $\qquad\qquad\qquad -E + 2A_0V\cos(\delta - \alpha_0) > 0$

or $\qquad\qquad\qquad V > \dfrac{E}{2A_0\cos(\delta - \alpha_0)}$

i.e. $\qquad\qquad V > \dfrac{E}{2A_0\cos\delta} \qquad$ (when α_0 is neglected) $\qquad (7.60)$

Expression (7.60) represents the condition of stability of the receiving-end voltage of a series capacitor compensated power transmission line.

If V_{cri} represents the critical receiving-end voltage at voltage stability limit, from Eq. (7.60)

$$V_{cri} = \dfrac{E}{2A_0\cos(\delta - \alpha_0)} \simeq \dfrac{E}{2A_0\cos\delta} \quad \text{(neglecting } \alpha_0\text{)} \quad (7.61)$$

Substituting Q_R from Eq. (7.52) in Eq. (7.49) and utilising Eq. (7.48), neglecting a_0, Eq. (7.49) finally becomes

$$V = \dfrac{E}{A_0}(\cos\delta - \sin\delta\tan\phi) \quad (7.62)$$

Comparing Eq. (7.58) with Eq. (7.62) at voltage stability limit and simplifying, we get

$$\dfrac{E}{2A_0\cos\delta} = \dfrac{E}{A_0}(\cos\delta - \sin\delta\tan\phi)$$

Further simplification yields

$$\delta_{lim} = \dfrac{\pi}{4} - \dfrac{\phi}{2} \quad (7.63)$$

where δ_{lim} represents the limiting power angle at the voltage stability limit.

For a series capacitor compensated line, the overall circuit constant A_0 is given by

$$A_0 = [(A_1 + X_c A_3)^2 + (A_2 + X_c A_4)^2]^{1/2} \quad (7.64)$$

when the capacitor is placed at the centre of the line, and A_0 is given by

$$A_0 = (a_1^2 + a_2^2)^{1/2} \tag{7.65}$$

when the capacitor is placed at the load end. Utilising Eq. (7.64) and Eq. (7.65) in Eq. (7.61), the receiving-end voltage at voltage stability limit is given by

$$V_{cri} = \frac{E}{2[(a_1^2 + a_2^2)]^{1/2} \cos \delta} \tag{7.66}$$

and

$$V_{cri} = \frac{E}{2[(A_1 + X_c A_3)^2 + (A_2 + X_c A_4)^2]^{1/2} \cos \delta} \tag{7.67}$$

for the capacitors placed at the load end and at the centre of the line respectively. The value of δ can be obtained from Eq. (7.63) for any power factor.

At unity power factor, from Eq. (7.66) or (7.67) and utilising Eq. (7.63),

$$V_{cri} = \frac{E}{\sqrt{2} \, A_0} \tag{7.68}$$

Comparison of Eq. (7.68) with Eq. (7.58) reveals that the critical receiving-end voltage at stability limit is governed by the condition of equality of the roots of the system voltage equation. Expressions denoted by Eq. (7.55) or Eq. (7.56) represent the critical amount of power transfer over the power line with series compensation at voltage stability limit.

Figure 7.19 represents the simulated results of the steady state receiving-end voltage versus the real power profile of a 220 kV, 275 km line at different power factors. The solid curves represent the profile of the receiving-end voltage for uncompensated lines, dotted curves for compensated lines when the series capacitors are placed at the receiving-end and chain dotted curves for the same compensated line, the series capacitor being placed at the centre of the line. It can be observed that the critical receiving-end voltage at voltage stability limit improves with the inclusion of series capacitors while the effect is best achieved when the capacitors are placed at the centre of the line.

With capacitors placed at the receiving-end, the generalised B_0 constant decreases in magnitude with compensation causing an improvement in voltage profile. For the same line, when the capacitors are placed at centre, the generalised constants A_0 and B_0 change in magnitude causing the voltage profile to improve further.

It can also be observed that the magnitude of V_{cri} at stability limit remains the same for uncompensated as well as for compensated lines where the capacitor is placed at load end.

Though the power transfer limit and the dynamic voltage profile increase remarkably using series compensation at load end, the critical value of the receiving-end voltage does not change with the load-end series compensation. However, for a compensated line when the series capacitor is placed at the middle, the situation is different where the voltage profile, maximum power transfer at

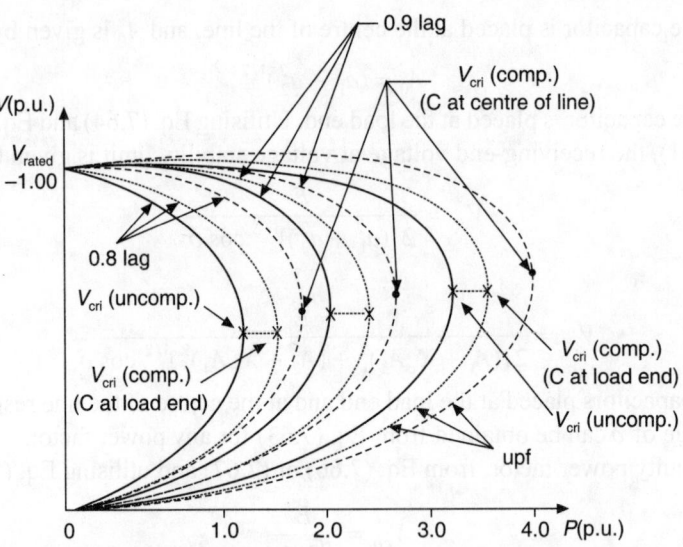

Figure 7.19 Profile showing V vs P in p.u. for uncompensated and series capacitor compensated line at different power factors.

stability limit as well as the receiving-end voltage at stability limit are improved. Reduction of the generalised constant A_0 serves to achieve this improvement.

A simulation has been performed to observe the profile of receiving-end voltage at stability limit (V_{cri}) against the series capacitive reactance (X_c) with ascending degree of compensation, where X_c decreases causing improvement in V_{cri} for the case when the series compensation is applied at the centre of the line. It is observed that the critical value of the receiving-end voltage does not vary with the degree of shunt compensation for the case when the capacitor is placed at the load end.

It has thus been revealed that the series capacitor compensation improves the receiving-end voltage profile as well as the maximum amount of real power transfer (Figure 7.20) at voltage stability limit. The critical voltage and power at voltage stability limit have also been determined analytically under this condition.

The location of series capacitors plays an important role in determining the magnitude of the critical voltage at stability limit. The placement of the series capacitor at the centre of line is extremely advantageous and most beneficial in order to raise the level of the receiving-end voltage at voltage stability limit as well as during the normal operating condition.

The degree of compensation has also been found to be effective in improving the magnitude of the critical receiving-end voltage at stability limit when the capacitor is placed at the centre of the line only. The simulation of the developed theory for a typical power transmission system establishes the supremacy of the installation of the capacitor at the centre of the line to enhance the critical power transfer and voltage stability.

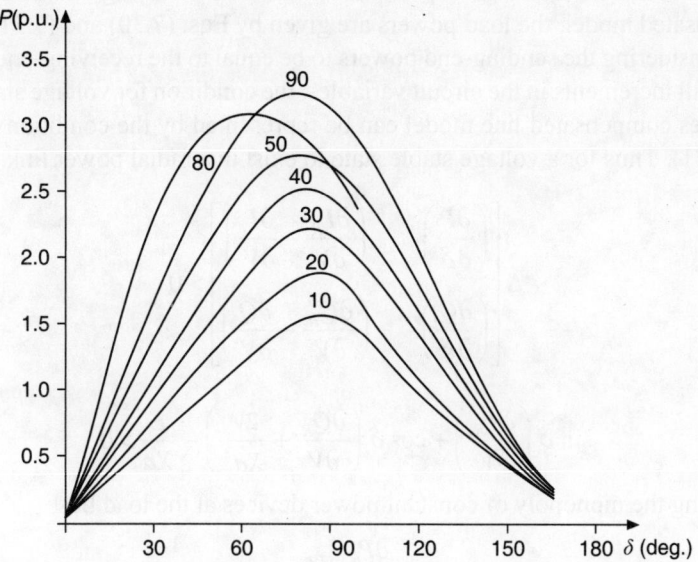

Figure 7.20 Profile of *P* vs *d* with varying series compensation.

7.8.6 Effect of Capacitor Bussing Arrangements and Contingency on Voltage Stability of Series Compensated Lines

This section presents a modified criterion of voltage stability of series capacitor compensated EHV lines for their operation in the voltage stable states utilising an existing method of approach and taking into account the voltage dependent load characteristics. Effects of series capacitor configurations and the selected outages have been studied simulating the EHV system model using a digital computer for the voltage instability.

Power flow equations in a lossless series compensated basic power transmission line model with a tap changing transformer may be given by

$$P = \frac{EV}{aX} \sin \delta = f_1(E, V, \delta) \qquad (7.69)$$

and

$$Q = \frac{EV}{aX} \cos \delta - \frac{V^2}{X^2 a^2} = f(E, V, \delta) \qquad (7.70)$$

where
  a = transformer off nominal tap ratio
  X = net line reactance = $(X_l - X_c)$ with
  X_l = inductive reactances in p.u.
  X_c = capacitive reactances in p.u.

Considering the voltage dependent loads at the receiving-end of the series compensated model, the load powers are given by Eqs. (7.30) and (7.31).

Considering the sending-end powers to be equal to the receiving-end powers and small increments in the circuit variables, the condition for voltage stability of the series compensated line model can be represented by the condition given in Eq. (7.71). Thus for a voltage stable state to exist in a radial power link,

$$\Delta \begin{bmatrix} \left(\dfrac{\partial P}{\partial \delta}\right) & \left(\dfrac{\partial P_R}{\partial V} - \dfrac{\partial P}{\partial V}\right) \\ \left(\dfrac{\partial Q}{\partial \delta}\right) & \left(\dfrac{\partial Q_R}{\partial V} - \dfrac{\partial Q}{\partial V}\right) \end{bmatrix} > 0 \qquad (7.71)$$

i.e. $$\sin \delta \left(\dfrac{\delta P_R}{\delta V}\right) + \cos \delta \left(\dfrac{\partial Q_R}{\partial V} + \dfrac{2V}{Xa^2}\right) - \dfrac{E}{Xa} > 0 \qquad (7.72)$$

Assuming the monopoly of constant power devices at the load bus,

$$\dfrac{\partial P_R}{\partial V} = 0 \qquad (7.73)$$

Thus, using Eq. (7.72) in Eq. (7.71) and simplifying yields

$$\cos \delta [(n P_R^0 \tan \phi V^{n-1}) + (2V/Xa^2)] - \dfrac{E}{Xa} > 0 \qquad (7.74)$$

Equation (7.74) represents the necessary condition for series capacitor compensated line to have a voltage stable state of operation. Denoting 'L' as an indicator of voltage stability of the series compensated power system, the comprehensive voltage stability indicator for series capacitor lines for different states may be defined as follows:

$L > 0$ for a stable voltage state,
$L = 0$ for the limiting stage of voltage stability
and $L < 0$ for the voltage unstable state.

Several arrangements of series capacitors can be made in the transmission circuits (Figure 7.21). Usually, the purpose of the capacitor insertion is to increase the steady state as well as the transient stability limit. However, with contingency, i.e. when a line section trips or is switched out for maintenance, the post contingency net transfer reactances change. Table 7.7 represents a comparison of the pre- and post-contingency transfer reactances of the capacitor arrangements shown in Figure 7.21.

A typical 400 kV, 400 km line has been simulated in the digital computer with lumped parameter concept and assuming the (r/x) ratio of the line to be very small. Pre-contingency moderately heavy loading has been assumed with steady state stability margin to be 25%. Post-contingency load has been assumed to be maintained in such a way so that the power angle between the sending-end and receiving-end buses does not change. The terminal reactances have been accounted in series with the line reactance and the source voltage has been taken as constant at its nominal value.

In the first stage, the magnitudes of L, the voltage stability indicator, have been computed for different types of capacitor connections of Figure 7.21 using Table 7.7 in the pre- and post-contingency period assuming voltage independent loads at the receiving-end bus when series compensation does not exist. Next, the different values of L have again been calculated and compared introducing series compensation, modelling the change in magnitude of series compensation by varying K in the pre- and post-contingency periods.

The same simulation has also been performed for the same system to observe the effect of link transformer tap ratio adjustment, receiving-end voltage magnitude variation, number of line sections, the load characteristic and the degree of series compensation on L and the voltage stability indicator in the post-contingency period.

The investigation reveals that the voltage stable states are with contingency, when moderately heavy line loading is desired to be maintained. The situation can be improved by series capacitor insertion though at a very high degree of compensation (found to be well above 80%), the system's voltage stability may again be lost. Different schemes of the series capacitor insertion have little effect in enhancing the voltage stable state as the magnitudes of L have been found to

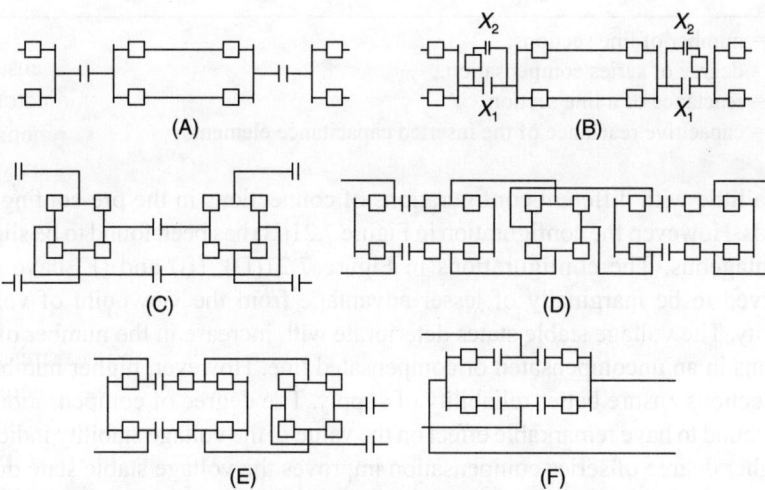

Figure 7.21 Several series capacitor arrangements.

A—Fixed series capacitors located in intermediate switching station bus
B—Switched series capacitors in same location as in A
C—Series capacitors in each line section and switched with line
D—Series capacitors incorporated into a breaker-and-a-half scheme in intermediate switching station bus. This is satisfactory for an odd number of intermediate switching stations
E—Same arrangement as in D, but showing the last intermediate switching station for an even number of stations
F—Split series capacitors, equivalent to D but with fewer breaker positions and higher vulnerability

Table 7.7 Pre- and post-contingency transfer reactances for different types of connections

Type of configuration	Net transfer reactance when	
	All lines are in service	One line section is out of service
A	$\dfrac{r}{2} x (1-k)$	$\dfrac{r}{2} x (1-k) + \dfrac{x}{2}$
B	$\dfrac{r}{2} x (1-k)$	$\dfrac{x}{2}[1 + r(1-k)] - \dfrac{x_1^2}{x_1 + x_2}$
C	$\dfrac{r}{2} x (1-k)$	$\dfrac{r+1}{2} x (1-k)$
D	$\dfrac{r}{2} x (1-k)$	$\dfrac{r}{2} x (1-k) + \dfrac{x}{2}(1-2k)$
E	$\dfrac{r}{2} x (1-k)$	$\dfrac{x}{2}\left[r + 1 - kr\dfrac{(r+3)}{(r+1)}\right]$
F	$\dfrac{r}{2} x (1-k)$	$\dfrac{r}{2} x (1-k) + \dfrac{x}{2}(1-2k)$

r = number of line sections.
k = degree of series compensation.
x = reactance of a line section.
x_i, x_2 = capacitive reactance of the inserted capacitance elements

vary a little with different configurations of connections in the pre-contingency periods. However, the configuration in Figure 7.21(B) has been found to be slightly advantageous. The configurations in Figure 7.21(D), (E) and (F) have been observed to be marginally of lesser advantage from the viewpoint of voltage stability. The voltage stable states deteriorate with increase in the number of line sections in an uncompensated or compensated line. However, higher number of line sections ensure better reliability of supply. The degree of compensation has been found to have remarkable effect on the value L, the voltage stability indicator. A higher degree of series compensation improves the voltage stable state during contingency till the system is over-compensated.

The initial power transfer being low, it is easier to maintain voltage stability of the system. The situation deteriorates with the drop in receiving-end voltage. The presence of voltage dependent loads further complicates the problem of voltage instability. Series compensation is effective in enhancement of voltage stability. Tap-changing transformers, do on the other hand, push the system towards voltage instability whenever they are in operation during contingency.

7.9 AN OPTIMAL LOAD SHEDDING SCHEME TO MAINTAIN VOLTAGE STABILITY FOLLOWING A LINE CONTINGENCY

7.9.1 Introduction

In the 'present day power system', the supply of uninterrupted electric power has become an important and necessary factor and it has become imperative for utilities to provide a reliable source of energy to the customers. In an emergency situation, where the integrity of a large power network is jeopardised, the operating personnel are often burdened with a large volume of data and are under pressure of taking immediate decisions. However, this decision making may not be at all effective unless operators can adequately ascertain the critical conditions that are prevalent on the network. These critical conditions can commonly be diagnosed by detecting the frequency and voltage problems in the network. In robust power systems, it is a usual practice to reroute power flows through alternate paths in case the system is threatened following a contingency. However, in the longitudinal power supply systems, the number of EHV lines are few and the .system is topographically radial; hence most of the time the operators are compelled to apply the last resort, i.e. load shedding, in order to save the power system, at least partially, from collapse while encountering a contingency.

It is believed that the best results for load shedding can be obtained if it is done locally. Also load shedding should not cause any further deterioration of the system by overloading the existing healthy lines and the degradation of service to the customers should be minimised for a given demand. The main objective of load shedding, in case the state of operation of the power system goes down to its critical level, is to keep the system operative after a contingency.

In the following section, with the above objective in mind, a simplified and straightforward load curtailment, approach has been presented in case the voltage stability of the power system is threatened following a contingency. The concept of load curtailment enables its application in quantifying the magnitude of the load curtailment, following a contingency or when the voltage stable state of the system is endangered.

Literature survey has revealed pioneer work in developing the concept of 'Performance Index (PI)' as well as the sensitivity of performance index in the field of contingency ranking and security assessment of power systems. A newer concept of 'Voltage Stability Performance Index (VSPI)' has been introduced in this chapter utilising the concept of 'Performance Index'developed by Medicherla et. al and has been applied to assess the voltage security of line model following a line contingency. Further, an experiment that has been conducted in a laboratory on a real time model power system has also been discussed to check the validity of the developed load shedding approach and to observe the governing effect of the voltage dependent loads, load power factor and magnitude of load curtailment.

In the preceeding chapters, analytical expressions for the critical receiving-end voltage and power angle of a shunt compensated and uncompensated power supply systems, at voltage stability limit, have been developed. Here, the effect of contingency on the receiving-end bus voltage has been presented taking into consideration the severest contingency that results in the decrease of the receiving-end voltage magnitude to the critical value. An optimal load shedding scheme is thus necessary in order to restore the receiving-end bus voltage from the critical value to the pre-contingency magnitude. The concept of 'Voltage Stability Performance Index (VSPI)' is helpful in order to compare the performances of power supply systems in encountering contingency from the viewpoint of voltage instability and effectiveness of shunt compensation.

7.9.2 Analysis

Assume the simplest reactive power-constrained EHV power supply system, operating satisfactorily at a given loading condition in the pre-contingency state. A contingency in the transmission system is assumed to be responsible for the receiving-end bus voltage to undergo a change, the amount of this change being governed by the severity of contingency. The bus voltage may even enter in the vicinity of voltage collapse in extreme cases and under this circumstance, the countermeasures may be of no use to restore the receiving-end voltage magnitude to the voltage stable state. In the lossless uncompensated line model, the steady state power flows are given by

$$P = \frac{EV_0^N}{X} \sin \delta \tag{7.75}$$

and

$$Q = \frac{EV_0^N}{X} \cos \delta - \frac{(V_0^N)^2}{X} \tag{7.76}$$

The determinant of the Jacobin of these power flow equations are given by

$$\Delta[J] = \frac{1}{X}[-E^2 V_0^N + 2EV_0^{N^2} \cos \delta] \tag{7.77}$$

Utilising the singularity of Jacobian, (i.e. when $\Delta|J| = 0$) at the voltage stability limit,

$$V^{cri} = \frac{E}{2 \cos \delta_{\lim}} \tag{7.78}$$

the limiting power angle ($\delta_{\lim}$) being given by

$$\delta_{\lim} = \frac{\pi}{4} - \frac{\phi}{2} \tag{7.79}$$

On the other hand, for a shunt capacitor compensated EHV line, analysing the Jacobian in the similar manner, the critical voltage and power angle of the compensated system are given by

$$V_{\text{comp}}^{\text{cri}} = \frac{E}{2\cos\delta_{\lim}(1 - x\cdot\omega c)} \tag{7.80}$$

and

$$\delta_{\lim(\text{comp})} = \frac{\pi}{4} - \frac{\phi}{2} \tag{7.81}$$

With contingency, the bus voltage drops in order to supply the pre-contingency loads and in the case where the generation of real and reactive powers is limited in the system (which is a very common phenomenon in an EHV radial transmission system), there is no other alternative to the operator than to shed the loads at the receiving-end bus. Voltage collapse being the ultimate fate of the receiving-end voltage dropping below the voltage stability limit, the assumption of the post-contingency voltage level at the voltage stability limit represents the severest contingency that requires maximum load shedding.

Again, the voltage excursion in the receiving-end bus following a contingency is given by

$$\Delta V = V_0^N - V^{\text{cri}} \tag{7.82}$$

Under the severest post-contingency condition, when the receiving-end bus voltage reaches its stability limit, the system can be represented as

$$[I_R] = [V^{\text{cri}}][Y + \Delta Y] \tag{7.83}$$

Utilising Eq. (7.82) and substituting the value of I_R, given by $\left[I_R = \dfrac{P - jQ}{V^{\text{cri}}} \right]$ in Eq. (7.83), the system equation can be represented as

$$(P - jQ) = (V_0^N - \Delta V)^2 Y + (V^{\text{cri}})^2 \Delta Y$$

$$= V_0^{N^2} \cdot Y + (\Delta V)^2 Y - 2\Delta V Y \cdot V_0^N + (V^{\text{cri}})^2 \Delta Y \tag{7.84}$$

However, in the pre-contingency state, the system can be represented as

$$V_0^{N^2} \cdot Y = P - jQ \tag{7.85}$$

which, on substitution in Eq. (7.84) yields,

$$\Delta Y = [2V_0^N \Delta V Y - (\Delta V)^2 Y]/(V^{\text{cri}})^2 = \frac{2\Delta V}{(V^{\text{cri}})^2} \cdot V_0^N Y$$

$$= \frac{2\Delta V}{(V^{\text{cri}})^2} \cdot \frac{(P - jQ)}{V_0^N} \quad \text{[utilising Eq. (7.85)]} \tag{7.86}$$

As the change in the admittance (i.e. the effect of contingency) is assumed to be the sole factor responsible for the drop in load bus voltage from V_0^N to V^{cri}, the optimum load to be curtailed (MLS), for securing the voltage stable state of operation, is given by

$$(\text{MLS}) = (V^{\text{cri}})^2 (\Delta Y)^*$$

$$= 2(P + jQ) \cdot \frac{\Delta V}{V_0^N} \qquad (7.87)$$

Hence, the real and reactive power load to be shedded (MLS_P and MLS_Q) can be expressed from Eq. (7.87) as

$$(MLS)_P = P \cdot \frac{2\Delta V}{V_0^N} = \Delta P, \quad \text{say} \qquad (7.88)$$

and

$$(MLS)_Q = Q \cdot \frac{2\Delta V}{V_0^N} = \Delta Q, \quad \text{say} \qquad (7.89)$$

Combining Eqs. (7.88) and (7.89), the net magnitude of the MVA power to be curtailed to ensure a voltage secured state of operation, is given by

$$(MLS) = \Delta S = [(\Delta P)^2 + (\Delta Q)^2]^{1/2} \qquad (7.90)$$

Utilising Eq. (7.82), Eqs. (7.88) and (7.89) can further be represented as

$$\Delta P = P \cdot 2 \left(1 - \frac{V^{cri}}{V_0^N} \right) \qquad (7.91a)$$

$$\Delta Q = Q \cdot 2 \left(1 - \frac{V^{cri}}{V_0^N} \right) \qquad (7.91b)$$

Assuming $V_0^N = E$ in uncompensated line model and utilising Eq. (7.78), Eqs. (7.90) and (7.91) become

$$\Delta P = P \cdot 2 \left(1 - \frac{1}{2 \cos \delta_{lim}} \right) \qquad (7.92)$$

$$\Delta Q = P \cdot 2 \left(1 - \frac{1}{2 \cos \delta_{lim}} \right) \qquad (7.93)$$

In a reactive power-constrained line, the critical power angles are governed by load power factor as shown by Eq. (7.79) and thus Eqs. (7.92) and (7.93) can also be represented in terms of load power factor as shown below in Eqs. (7.94) and (7.95), i.e.

$$\Delta P = 2P \left[1 - \frac{1}{2.0 \cos(\pi/4 - \phi/2)} \right] \qquad (7.94)$$

and

$$\Delta Q = 2Q \left[1 - \frac{1}{2.0 \cos(\pi/4 - \phi/2)} \right] \qquad (7.95)$$

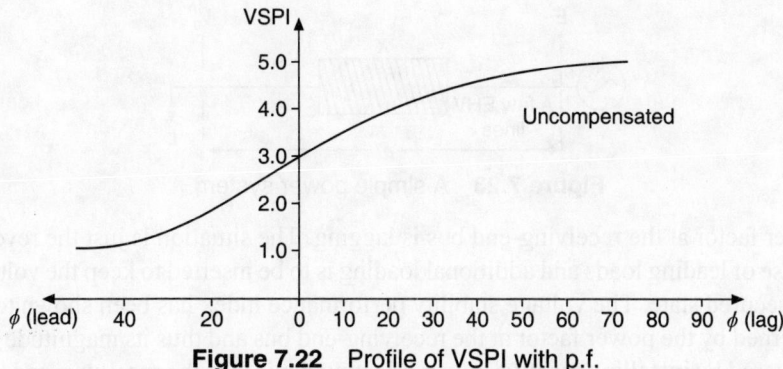

Figure 7.22 Profile of VSPI with p.f.

By a similar approach, for the shunt capacitor compensated line, the expressions for ΔP and ΔQ can be developed as

$$\Delta P = 2P\left(1 - \frac{1}{2\cos\delta_{\lim}(1 - x_{new}\cdot\omega c)}\right) \quad (7.96)$$

also

$$\Delta P = 2P\left(1 - \frac{1}{2[\cos(\pi/4 - \phi^N/2)][1 - x_{new}\cdot\omega c]}\right) \quad (7.97)$$

and

$$\Delta Q = 2Q\left(1 - \frac{1}{2\cos\delta_{\lim}(1 - x_{new}\cdot\omega c)}\right) \quad (7.98)$$

also

$$\Delta Q = 2Q\left(1 - \frac{1}{2[\cos(\pi/4 - \phi^N/2)][1 - x_{new}\cdot\omega c]}\right) \quad (7.99)$$

A concept of 'Voltage Stability Performance Index' (VSPI) can here be introduced and defined as

$$\text{VSPI} = [(\Delta P)^2 + (\Delta Q)^2]^{1/2} \quad (7.100)$$

where ΔP and ΔQ are given by Eq. (7.96) or (7.97) and Eq. (7.98) or (7.99) respectively.

Figure 7.22 represents the profile of the computed voltage stability performance index (VSPI) for varying power factor in the post-contingency period, for a specific amount of pre-contingency power flow through a radial EHV power link connecting a distant source of generation to a load centre (Figure 7.23). Each point on this profile indicates the optimum amount of load to be curtailed so as to keep the system at a voltage stable state following a contingency in the transmission system. This would normally have caused the receiving-end voltage to drop down to the voltage stability limit in the absence of the corrective measure, when the

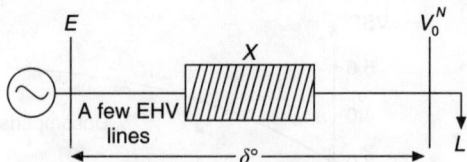

Figure 7.23 A simple power system.

power factor at the receiving-end bus is lagging. The situation is just the reverse in case of leading loads and additional loading is to be inserted to keep the voltage at a secured state. The voltage stability performance index has been shown to be governed by the power factor at the receiving-end bus and thus its magnitude can be altered by installing reactive power injecting devices at the receiving-end bus. Figure 7.24 represents such a typical simulation which reveals that the amount of load to be curtailed in order to upgrade the voltage of the receiving-end bus from its critical value to the rated value in the post-contingency period could remarkably be reduced incorporating the reactive power injecting device (shunt capacitors in the present case) at the receiving-end bus.

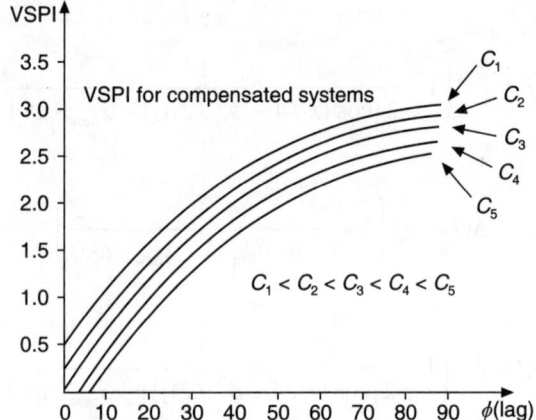

Figure 7.24 VSPI profile vs p.f. angle with compensation.

7.10 SYNCHRONOUS CONDENSER AT THE LOAD BUS

Modelling of synchronous condenser

A synchronous condenser is a synchronous machine designed for shunt reactive power compensation. In steady state operation, with a fixed excitation, the synchronous condenser can be represented by a generated emf V in series with the synchronous reactance (X_{ds}) of its own (Figure 7.25(a)). The operating V/I characteristics are shown in Figure 7.25(b). Each line in this figure represents a fixed value of field current (I_f) or the equivalent voltage V_0. These characteristics are applicable only when the field current is fixed and the machine is thus operating under the maximum excitation limit.

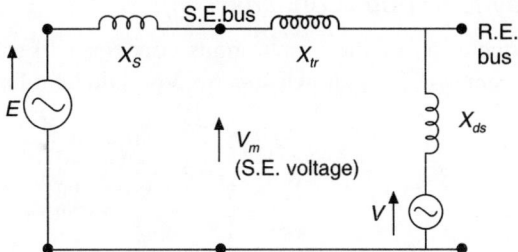

Figure 7.25(a) Representation of synchronous condenser
X_S = source reactance, X_{tr} = line reactance
X_{ds} = machine reactance, V = machine emt.

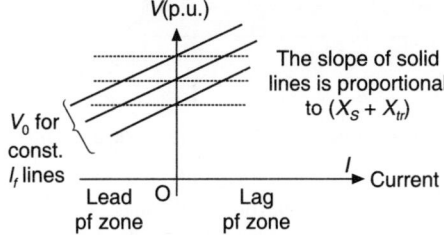

Figure 7.25(b) Operating characteristic (steady state) of a synchronous condenser.

In the transient state, the synchronous condenser behaves in the same way as that of steady state, however, the synchronous reactance being replaced by transient reactance X'_{ds}. The magnitude of X'_{ds} is much smaller than X_{ds}. Thus in the transient voltage current characteristic, (shown in Figure 7.26(a)), the curves are more flat, the slopes being proportional to $(X'_{ds} + X_{tr})$. The equivalent circuit of the machine at transient state of operation has been shown in Figure 7.26(b).

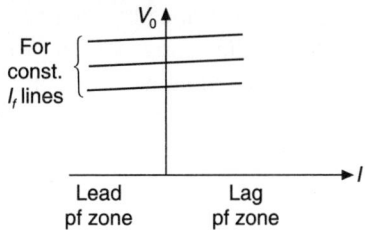

Figure 7.26(a) Transient state operating characteristic of synchronous condenser.

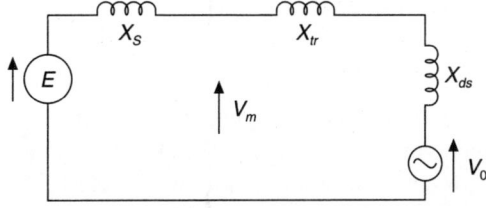

Figure 7.26(b) Representation of synchronous condenser in transient state.

Operation of synchronous condenser

To explain the operation of the synchronous condenser in transient state, the transient V–I characteristic is redrawn superimposed on load lines (Figure 7.27).

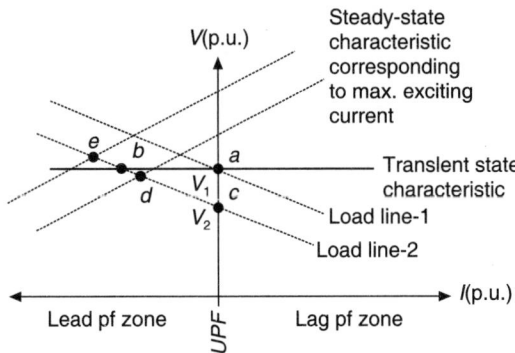

Figure 7.27 V–I characteristics of the synchronous condenser with load lines.

Assuming the shifting of load level, the load line-1 shifts to load line-2. If there is no condenser, the system voltage would have been dropped from point 'a' to point 'c', i.e. from V_1 to V_2. However, the presence of the condenser reacts to raise the voltage again near to the nominal value as is evident by the intersection of the load line-2 with the transient state characteristic (point 'b'). The machine is to operate in the lead p.f. zone.

Thus the new operating point is the point 'b' where the condenser is operating in the over-excited stage. However, if the capacity of the condenser at this operating point is beyond the rated capacity of the condenser, the operation at this point is limited for a short interval of time depending on the machines' overload capability. Then the field current will be automatically reduced by adjusting the voltage regulator and a new operating point corresponding to the nearest steady-state characteristic (at maximum excitation limit) will be attained at point 'd'.

On the other hand, if the operation of the condenser corresponding to the transient state of operation point 'b' is within the capability limit, the steady-state operation is then settled at point 'e' of the nearest steady-state operating characteristic.

Application of synchronous condensers to have limited control of voltage at the receiving-end

Here, a constant reactive power source (a synchronous condenser at a fixed excitation limit) with its VAR output Q_{max} is considered to be added at the receiving-end in order to increase the power transfer, enhance the stability limits and regulate the receiving-end voltage profile keeping it well within the design limit allowing maximum voltage regulation of 5% (i.e. $V_{min} = 0.95$ p.u.).

With the enhancement of VAR reserves, the operating point voltage and stability margin improve. The capability of this VAR reserve with the complete utilisation of the desired stability margin (M_d) is given by

$$Q_{max} = \frac{V_d^2}{X} - \frac{E_d}{X} \cos \delta_d \qquad (7.101)$$

where

$$\delta_d = \sin^{-1}\left(\frac{P_{Rd} \cdot X}{V_d \cdot E}\right) \qquad (7.102)$$

and

$$P_{Rd} = P_{Ro} \cdot \frac{1}{1 - M_d} \qquad (7.103)$$

Further increase in the real power transfer through the tie is possible till the VAR supply capability at the receiving-end is totally utilised and the receiving-end bus voltage is maintained at 0.95 p.u. At unity power factor under this condition,

$$Q_R = \frac{EV_{cri}}{X} \cdot \cos \delta_{cri} - \frac{V_{cri}^2}{X} + Q_{max}$$

which, under the operating condition of unity power factor at load bus, can be represented as

$$Q_{max} = \frac{V_{cri}^2}{X} - \frac{EV_{cri}}{X} \cdot \cos \delta_{cri}$$

[$Q_R = 0$ under this condition].
Thus,

$$\delta_{cri} = \cos^{-1}\left(\frac{V_{cri}}{E} - \frac{Q_{max} \cdot X}{EV_{cri}}\right) \qquad (7.104)$$

and

$$P_{cri} = \frac{EV_{cri}}{X} \sin \delta_{cri} \qquad (7.105)$$

Real power transfer may further be increased at the cost of receiving-end bus voltage till the steady state stability limit is attained. When the reactive power source acts as a constant VAR supply and the receiving-end voltage again becomes uncontrolled, at unity power factor, the receiving-end bus voltage at steady state power limit is obtained from

$$Q = \frac{EV_{lim}}{X} \cos \delta_{lim} - \frac{V_{lim}^2}{X} + Q_{max}$$

i.e.

$$V_{R_{\lim}} = \frac{1}{2}(E\cos\delta_{\lim} + \sqrt{E^2\cos^2\delta_{\lim} + 4Q_{\max}X})\qquad(7.106)$$

The limiting power angle at steady state stability limit is given by

$$\delta_{\lim} = \cos^{-1}\left(\frac{E}{\sqrt{2E^2 + 4Q_{\max}X}}\right)\qquad(7.107)$$

when the power limit is expressed as

$$P_{R_{\lim}} = \frac{EV_{\lim}}{X}\sin\delta_{\lim}\qquad(7.108)$$

Analysis with real system data of a tie line clearly highlights the utility of the reactive power source at the receiving-end for enhancement in power flow and voltage reliability, the results being presented in Tables 7.8(a) and 7.8(b).

Table 7.8(a) Computational results of operating variables of a typical tie line at specific limits

(a) Receiving-end bus voltage uncontrolled:

$P_{R_{\max}}$ (p.u.)	$\delta_{\max}$ (°)	$V_{\max}$ (p.u.)
4.998	45.000	0.707

Table 7.8(b) Limited voltage control at receiving-end bus with initial operating point

$P_{R0} = 5$ p.u. (0.5 SIL), $M_0 = 30\%$ and $E = 1.00$ p.u.

P_{R_d} (p.u.)	δ_d (0)	$Q_{\max}$ (p.u.)	$P_{R_{cri}}$ (p.u)	δ_{cri} (0)	$P_{R_{\lim}}$ (p.u.)	$\delta_{\lim}$ (0)	$V_{\lim}$ (P.u.)
7.140	48.753	3.240	7.540	52.492	7.582	56.570	0.908

7.11 FACTS DEVICES

FACTS (*Flexible AC Transmission System*) is defined as an alternating current transmission system incorporating semiconductor-based power electronic and other static controllers in order to enhance the power transfer capacity and increase the controllability of transmission. It increases the ability to accommodate changes in operating conditions simultaneously maintaining steady state and transient stability margins.

FACTS devices include power electronic and other static devices with advanced power conversion and switching capabilities. The most dominant converters needed in FACTS controllers (the terms *device* as well as *controllers* will be used synonymously throughout the text as FACTS have power electronic devices being used as controllers) are the voltage source converters, based on devices with gate turn-off capability. In such unidirectional voltage converters, the power reversal involves reversal of current and not the voltage. A number of FACTS controllers need energy *storage* arrangements and use capacitors, batteries and superconducting magnets as *storage devices*. The addition of a storage device makes FACTS controller more efficient in controlling the system dynamics. Dynamic pumping of real power is possible with storage device enabled FACTS controllers.

Before entering into the arena of a brief description of a variety of FACTS devices, it is worth mentioning here that for the converter-based controllers, there are two principal types of converters with gate turn-off devices. These are popularly voltage source converters and current source converters. In a voltage source converter, unidirectional dc voltage of a dc capacitor is introduced on ac side as ac voltage using a proper switching technology. It is then possible to vary the ac output voltage in magnitude and also in phase compared to ac system voltage. The power reversal involves reversal of current, not the voltage. When the storage capacity of the unit is small and there is no other power source connected to it, the converter unit cannot inject or absorb real power beyond a short period. The ac output voltage is maintained at ($\pi/2$) leading or lagging the current on ac side and thus the converter is used to either absorb or supply the reactive power only.

In case the FACTS technology uses the current source converter, the dc current is passed to the ac side through the switching of devices as ac current. This switching-based ac current is variable in magnitude and phase with respect to the ac system voltage. The power reversal involves reversal of voltage and not current. In FACTS technology, from the point of view of cost factor, the voltage source converters are preferred over the current source converters.

FACTS controllers may use thyristor devices with no gate turn-off capability or may use power devices with gate turn-off capability. Most of the controllers with gate turn-off capability are the dc to ac converters which can exchange active and/or reactive power with the ac bus at which the FACTS controller is connected. When the task of the FACTS controller is to control the reactive power only, it is provided with minimum storage on the dc side. However, if the need is to generate ac voltage or current with more than 90° phase difference with respect to the corresponding quantities of ac system, the storage device need augmentation using capacitors, batteries and superconducting magnets.

7.12 CLASSIFICATION OF FACTS CONTROLLERS

FACTS controllers can generally be divided into four categories:
  (a) Series controllers
  (b) Shunt controllers
  (c) Series-series controllers
  (d) Series-shunt controllers.

7.13 SERIES CONTROLLERS

Series controllers, in FACTS technology, are used to inject voltage in series with the line. In its simplest form, a variable impedance multiplied by the current flow through it represents an injected series voltage in the line. If the series voltage is in-phase quadrature with the line current, the series controller only supplies or absorbs variable reactive power. Real power is involved for any other phase relationship between the injected voltage and the line current. The symbolic representation of the series FACTS controller is shown in Figure 7.28.

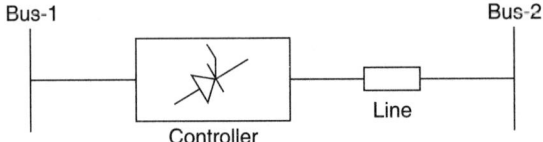

Figure 7.28 Representation of series FACTS controller.

Series controllers have direct impact on the voltage and hence the currents, and this helps in controlling the power flow directly using series controllers. Series controllers thus can control power flow in the line and can damp oscillations. The effectiveness of series controllers is more in compared to that of shunt controllers.

Series controllers are useful in controlling the series voltage drop in the line as these types of controllers can effectively inject (add or subtract) voltage in series. However, for several lines terminated in a bus, there would be a requirement of several series controllers. Moreover, they need to be protected from the possibility of abnormal transient voltage rise due to through faults in the line and need to be designed to encounter contingency and dynamic overloads.

7.14 TYPE OF SERIES CONNECTED CONTROLLERS

7.14.1 SSSC

SSSC (or S³C) is one of the most important FACTS series controllers. The full form is *Static Synchronous Series Compensator*. It is basically a static synchronous generator operated without an electrical energy source. It acts as a series compensator whose output voltage is in quadrature with the line current for the sake of increasing or reducing the overall reactive voltage drop in the line and thereby controlling the transmitted power. It is operated using a voltage-source converter or a current-source converter. The symbolic representation of the voltage-source converter based SSSC is shown in Figure 7.29. The transformer ratio is chosen in consonance with the economical design. If an extra energy source is added (Figure 7.30), SSSC can also be used to control the real power flow as it is possible to inject the series voltage with variable voltage angle. Otherwise, without any energy storage device this FACTS device can be used to control the reactive power only and can inject the variable voltage (90° leading or lagging the current). It is to be noted here that for SSSC, the primary of the transformer (and hence the secondary and the converters) has to carry full-load current including the fault current unless the converter is temporarily bypassed during faults.

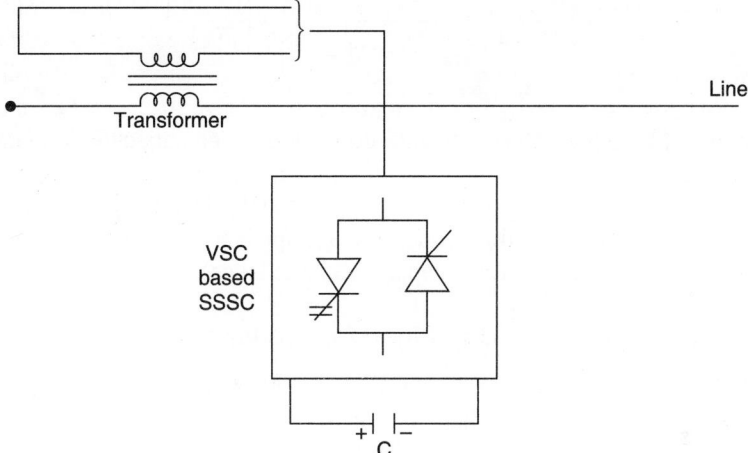

Figure 7.29 Symbolic representation of SSSC.

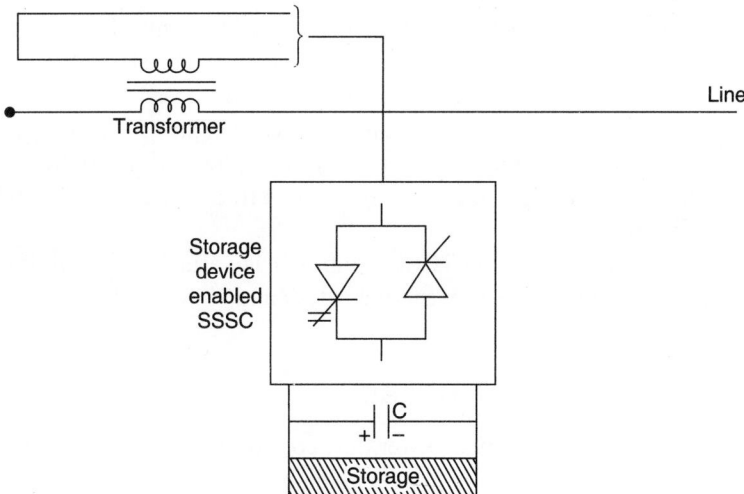

Figure 7.30 Storage device enabled SSSC connected to line.

The next variety of series-connected FACTS devices is TCSC (*Thyristor Controlled Series Capacitor*). In this device, a capacitive series reactance is shunted by a thyristor controlled reactor (TCR) in order to provide stepless variation in capacitive reactance (Figure 7.31).

TCSC is a very important FACTS device like SSSC. TCSC does not require thyristors with gate turn-off capability. In this scheme, the firing angles of the antiparallel thyristors are controlled to control the reactor and at 180° firing angle, the thyristor-controlled reactor is non-conducting while the series capacitor has its normal impedance. When the firing angle is reduced from 180° (but not less

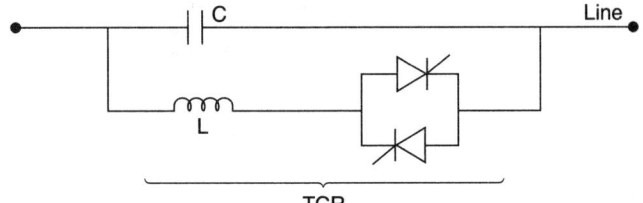

Figure 7.31 Schematic of thyristor-controlled series capacitor (TCSC).

than 90°), the capacitive impedance increases and at 90° firing angle, the thyristor controlled reactor conducts fully and the total impedance become inductive (the reactor impedance is less than the capacitive impedance).

7.14.2 TSSC and TCSC Combined Controller

TSSC (*Thyristor Switched Series Capacitor*) is another type of series compensator using the FACTS technology where a capacitive series reactance is shunted by a thyristor switched reactor to provide a step-controlled series capacitive reactance. Instead of continuous control of capacitive impedance, the control of inductance by switching gives better economy.

In a hybrid module of FACTS series controllers, TCSC may be clubbed with TSSSC such that in TCSC, the firing angle is changed for stepless control of impedance from inductive to capacitive range while TSSSC can be used for stepped controls. The TCSC also varies with the electrical length of the compensated transmission line and may be used to provide fast power flow control. It also increases the stability margin of the system and is effective in damping SSR and power oscillations.

Since the impedance of the TCSC controller is variable, we can write the power flow equation in a line between the sides i and j having TCSC controllers as discussed in the succeeding paragraphs.

7.14.3 Power Flow Model of TCSC

The TCSC can also serve as a real power flow controller between two nodes i and j. The expression of power flow $P_{ij(\text{reg})}$ is given by

$$P_{ij(\text{reg})} = \frac{|V_i||V_j|}{X_{(\text{TCSC})}} \sin(\delta_i - \delta_j) \tag{7.109}$$

where $X_{(\text{TCSC})}$ is the equivalent reactance of the TCSC controller which may be adjusted to regulate the active power flow across TCSC.

For inductive operation, the reactive power equation at node i is

$$Q_i = \frac{|V_i|^2}{X_{(\text{TCSC})}} - \frac{|V_i||V_j|}{X_{(\text{TCSC})}} \cos(\delta_i - \delta_j) \tag{7.110}$$

For capacitive operation, the signs of the equation are reversed. If we are to consider the power equations at node j, we can write

$$Q_j = \frac{|V_j|^2}{X_{(TCSC)}} - \frac{|V_j||V_i|}{X_{(TCSC)}} \cos(\delta_j - \delta_i) \qquad (7.111)$$

for inductive operation. For capacitive operation, the signs of the equation (7.111) are again reversed. For power flow studies, we can incorporate the TCSC controller in the formulations and we can write down the power mismatch equation considering the case when the power transfer is taking place from nodes i to j at a value P_{ij}. The equations can be oriented in matrix form as follows:

$$\begin{bmatrix} \Delta P_i \\ \Delta P_j \\ \Delta Q_i \\ \Delta Q_j \\ \Delta P_{ij} \end{bmatrix} = \begin{bmatrix} \frac{\partial P_i}{\partial \delta_i} & \frac{\partial P_i}{\partial \delta_j} & \frac{\partial P_i}{\partial |V_i|} & \frac{\partial P_i}{\partial |V_j|} & \frac{\partial P_i}{\partial X_{(TCSC)}} \\ \frac{\partial P_j}{\partial \delta_i} & \frac{\partial P_j}{\partial \delta_j} & \frac{\partial P_j}{\partial |V_i|} & \frac{\partial P_j}{\partial |V_j|} & \frac{\partial P_j}{\partial X_{(TCSC)}} \\ \frac{\partial Q_i}{\partial \delta_i} & \frac{\partial Q_i}{\partial \delta_j} & \frac{\partial Q_i}{\partial |V_i|} & \frac{\partial Q_i}{\partial |V_j|} & \frac{\partial Q_i}{\partial X_{(TCSC)}} \\ \frac{\partial Q_j}{\partial \delta_i} & \frac{\partial Q_j}{\partial \delta_j} & \frac{\partial Q_j}{\partial |V_i|} & \frac{\partial Q_j}{\partial |V_j|} & \frac{\partial Q_j}{\partial X_{(TCSC)}} \\ \frac{\partial P_{ij}}{\partial \delta_i} & \frac{\partial P_{ij}}{\partial \delta_j} & \frac{\partial P_{ij}}{\partial |V_i|} & \frac{\partial P_{ij}}{\partial |V_j|} & \frac{\partial P_{ij}}{\partial X_{(TCSC)}} \end{bmatrix} \begin{bmatrix} \Delta \delta_i \\ \Delta \delta_j \\ \Delta |V_i| \\ \Delta |V_j| \\ \Delta X_{(TCSC)} \end{bmatrix}$$

(7.112)

The real power flow mismatch equation in TCSC is given by

$$\Delta P_{ij} = P_{ij(reg)} - P_{ij(calculated)} \qquad (7.113)$$

The value of $X_{(TCSC)}$ of the TCSC controller can be updated at the end of the iteration p using the following relation

$$X_{(TCSC)}^{(p+1)} = X_{(TCSC)}^{(p)} + \Delta X_{(TCSC)}^{(p)} + X_{(TCSC)}^{(p)} \qquad (7.114)$$

7.14.4 TCSR

In the next stage we will see another FACTS technology-based series controller namely TCSR (*Thyristor Controlled Series Reactor*). It is a variable inductive reactance compensator and consists of a series reactor shunted by a thyristor-controlled reactor (Figure 7.32). It offers a smoothly variable series inductive reactance.

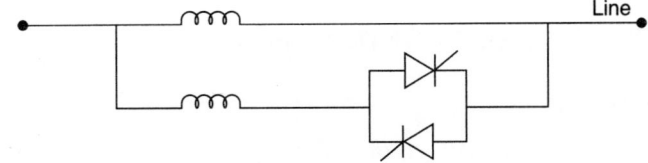

Figure 7.32 Schematic of thyristor-controlled series reactor (TCSR).

At firing angle of 180°, the TCR stops conducting and the series reactor acts as a fault limiter only. As the angle of trigger reduces from 180°, the network inductance decreases until the firing angle approaches 90°. At 90° firing angle the equivalent inductance is the parallel combination of the two inductors. Another variety of TCSR is TSSR (*Thyristor Switched Series Reactor*) which is also a series FACTS device and consists of a series reactor shunted by a thyristor-switched reactor in order to provide stepwise control of the series inductive reactance. It is a complement of TCSR but with thyristor switches either fully on or off in order to achieve a combination of stepped series inductance.

7.15 SHUNT CONTROLLERS

Similar to series controllers, the *shunt controllers* have variable impedance or a variable source or a combination of both. All shunt-connected FACTS devices inject current into the bus at the point of connection. The shunt impedance may be variable to vary the injected current. As long as this injected current is in phase quadrature with line voltage, the shunt controller only supplies or absorbs variable reactive power. Any other phase relationship of the generated current with line voltage will involve real power flow.

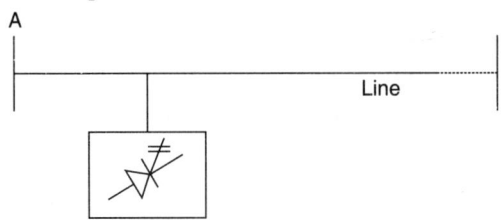

Figure 7.33 Symbolic representation of shunt controller.

A shunt controller (symbolically represented in Figure 7.33) is like a current source which draws from or injects currents in the bus. The injected current may be leading or lagging and this device is thus able to inject or absorb VAR alone or in combination of active and reactive currents. Also this device may be installed at the desired bus at which the controllability needs to be enhanced. Shunt type FACTS devices are very effective in voltage control and damping of voltage oscillations. The shunt controller is beneficial in controlling the voltage of a node at which several lines are terminated and is a cost-effective device for voltage control within a reasonable range. The shunt controllers can accommodate storage devices like series controllers to also inject real power to the connected node and hence adds a new dimension in their applicability.

7.16 TYPES OF SHUNT CONTROLLERS

7.16.1 SVC

Static VAR Compensator (SVC) is one of the most important shunt controllers in FACTS technology. It is a shunt connected static VAR generator (or absorber) whose output is designed to draw capacitive or inductive current so as to maintain normal voltage at the bus with which the SVC is connected.

SVC is based on thyristors without the gate turn-off capability. It includes a *thyristor controlled reactor* (TCR) in parallel with fixed capacitors in its commonest form and is known as FC TCR (*Fixed Capacitor Thyristor Controlled Reactor*) type SVC. For better control, the fixed capacitor itself can be replaced by a thyristor controlled capacitor (TSC) and the combination then becomes TSC TCR (*Thyristor Switched Capacitor Thyristor Controlled Reactor*). The TCR (*Thyristor Controlled Reactor*) or a TSR (*Thyristor Switched Reactor*) are also FACTS devices but may be treated as subsets of SVC. In TCR the firing angle is continuously variable while for TSR the thyristors act just as switches and there is no firing angle control. Similarly the TSC (*Thyristor Switched Capacitor*) is also a FACTS device and is obviously a subset of the SVC. The capacitors are switched in or out (without any firing angle control) as per requirement. Unlike shunt reactors, shunt capacitors cannot be switched continuously with variable firing angle control.

7.16.2 SSG

Static Synchronous Generator (SSG) is also a shunt controlled FACTS controller. It is a static self-commutated switching power device supplied from an electric energy source. It is required to generate a set of adjustable multi-phase output voltages, which may be coupled to an ac power network for the purpose of controlling the real and reactive power of the bus at which it is connected. SSG converts any source of electrical energy coming out from a battery, dc capacitor or any other dc source to ac voltage and current using chopper and converter technology. Different types of energy storage devices may form a section/part of SSG. A chemical-based energy storage system, known as *Battery Energy Storage System* (BESS) or *Superconducting Magnetic Energy Storage System* (SMESS), may be used as energy storage device in shunt controllers. They are all FACTS devices and subsets of different types of FACTS devices.

7.16.3 Power Flow Model of SVC

The simplest form of SVC model represents it as a generator with zero active power output but having reactive generation with reactive power limits. As long as the SVC is in operation, the bus with which it is connected act as a PV bus.

A simple and effective way to include the SVC in power flow techniques is to use this device as a variable susceptance. The shunt susceptance represents the total SVC susceptance necessary to maintain the voltage magnitude at the bus at the specified value. Inclusion of SVC at any load node makes that node the voltage-controlled node and at this node the voltage magnitude and the nodal active and reactive power are specified; the variable susceptance B_{SVC} is operated as a state variable. The philosophy is that if B_{SVC} is within limits, the specified voltage is attained and the bus operates as PV bus. When B_{SVC} is above or lower than the limits, B_{SVC} becomes fixed at the violated limit and the node becomes the PQ bus again if there is no other voltage regulating equipment present. The active and reactive power drawn by the SVC connected at node j are given by

$$P_j = 0; \quad Q_j = -|V_j|^2 B_{SVC}$$

Also the mismatches are given as

$$\begin{bmatrix} \Delta P_j \\ \Delta Q_j \end{bmatrix} = \begin{bmatrix} 0 & 0 \\ 0 & \partial Q_j / \partial B_{SVC} \end{bmatrix} \begin{bmatrix} \Delta \delta_j \\ \Delta B_{SVC} \end{bmatrix} \quad (7.115)$$

At the end of iteration p, the variable shunt susceptance is corrected as

$$B_{SVC}^{(p+1)} = B_{SVC}^{(p)} + \Delta B_{SVC}^{(p)} \quad (7.116)$$

which may further be written as

$$B_{SVC}^{(p+1)} = B_{SVC}^{(p)} + (\Delta B_{SVC} / B_{SVC})^{(p+1)} B_{SVC}^{(p)} \quad (7.117)$$

The changing susceptance represents the total SVC susceptance necessary to maintain the nodal voltage magnitude at the specified value. SVC compensation may also be computed in terms of the thyristor firing angle. However, the additional calculation requires an iterative solution as the SVC susceptance and thyristor firing angle are non-linearly related.

The steady-state susceptance of SVC can be obtained from the relation,

$$B_{SVC} = B_c - B_{TCR} = \frac{1}{X_c X_\ell} \left[X_\ell - \frac{X_c}{\pi} \{ 2(\pi - \alpha) + \sin(2\alpha) \} \right]$$

where $X_\ell = \omega L$ and $X_c = \dfrac{1}{\omega C}$.

Since, $Q_{SVC} = -V_j^2 B_{SVC}$, we can write

$$Q_j = -\frac{V_j^2}{X_c X_\ell} \left[X_\ell - \frac{X_c}{\pi} \{ 2(\pi - \alpha) + \sin(2\alpha) \} \right]$$

(assuming that Q_j is the jth bus reactive power injection due to SVC installation at the jth bus). The linearised SVC equation is then given by

$$\begin{bmatrix} P_j \\ Q_j \end{bmatrix}^{(p+1)} = \begin{bmatrix} 0 & 0 \\ 0 & \dfrac{2V_j^2}{\pi X_\ell} [\cos(2\alpha) - 1] \end{bmatrix}^{(p+1)} \begin{bmatrix} \Delta \delta_j \\ \Delta \alpha \end{bmatrix}^{(p+1)} \quad (7.118)$$

α being the firing angle of SVC. Therefore, at the end of iteration, the variable firing angle α is updated by the equation,

$$\alpha^{p+1} = \alpha^p + \Delta \alpha^p \quad (7.119)$$

The bus with which the SVC is connected (i.e. the jth bus) becomes a voltage-controlled bus where the voltage magnitude and active and reactive powers are specified along with either the SVC firing angle (α) or the SVC equivalent susceptance (B_{SVC}). If α or B_{SVC} is within limits, the specified voltage magnitude is attained and the bus remains as voltage controlled (PV) bus. However, if α or B_{SVC} go beyond limits, then these variables are fixed at the violated limits and the bus becomes a pure load (PQ) bus.

7.16.4 STATCOM

Static Synchronous Compensator (STATCOM) is another shunt connected FACTS device. It is a static synchronous generator operated as a static VAR compensator whose capacitive or inductive output currents are controlled to control the bus voltage with which it is connected.

STATCOM operation is based on the principle of voltage source or current source converter. The schematic of a STATCOM is shown Figure 7.34.

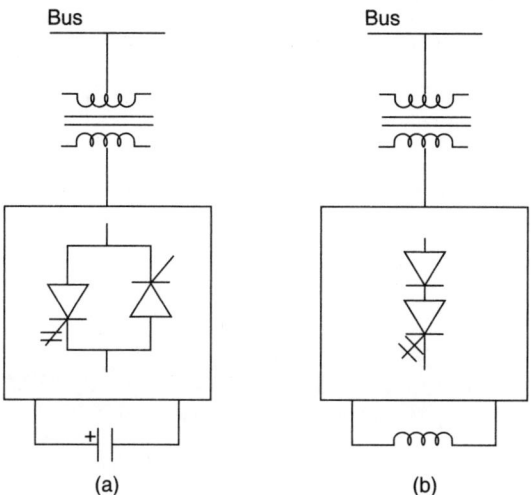

Figure 7.34 STATCOM schematic: (a) with voltage source converter, (b) with current source converter.

The STATCOM goes on well with energy storage facilities. When used with voltage-source converter, its ac output voltage is controlled such that the required reactive power flow can be controlled at the load bus with which it is connected. Due to the presence of dc voltage source in the capacitor, the voltage-source converter converts its voltage to ac voltage source and controls the bus voltage. STATCOM can also be designed to act as an active filter to reduce system harmonics and frequently includes the facility of having active power control.

7.16.5 Power Flow Model of STATCOM

Suppose,

$V_j \angle \delta_j$ = bus voltage at bus j

$V_{sc} \angle \delta_{sc}$ = inverted voltage (ac) at the output of STATCOM, referred to the jth bus side
X_{sc} = STATCOM reactance
Q_{sc} = reactive power exchange for the STATCOM with the bus.

Obviously,

$$Q_{sc} = \frac{|V_j|^2}{X_{sc}} - \frac{|V_j||V_{sc}|}{X_{sc}} \cos(\delta_j - \delta_{sc}) \quad (7.120)$$

$$= \frac{|V_j|^2 - |V_j||V_{sc}|}{X_{sc}}, \text{ if } \delta_j = \delta_{sc} \text{ (for a lossless STATCOM).}$$

Thus if $|V_j| < |V_{sc}|$, Q_{sc} becomes negative and the STATCOM generates reactive power. On the other hand, if $|V_j| > |V_{sc}|$, Q_{sc} becomes positive and the STATCOM absorbs reactive power.

Also,
$$V_{sc} = |V_{sc}|(\cos\delta_{sc} + j\sin\delta_{sc}) \quad (7.121)$$

The maximum and minimum limits of $|V_{sc}|$ will be governed by the STATCOM capacitor rating. δ_{sc} may have any value between 0° to 180° but usually δ_{sc} is kept nearly equal to δ_j.

Let us now draw the equivalent circuit of STATCOM (Figure 7.35).

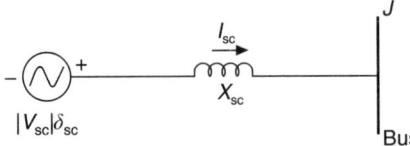

Figure 7.35 Equivalent circuit of shunt-operated STATCOM.

Here,
$$I_{sc} = Y_{sc}(V_{sc} - V_j)$$

where
$$Y_{sc} = \frac{1}{Z_{sc}} = G_{sc} + jB_{sc}$$

∴ S_{sc} (complex power flow) $= V_{sc} I_{sc}^* = V_{sc} Y_{sc}^* (V_{sc}^* - V_j^*) \quad (7.122)$

However, $V_{sc} = |V_{sc}|(\cos\delta_{sc} + j\sin\delta_{sc})$ [Eq. (7.121)]

Substitution of V_{sc} in expression of S_{sc} leads to the following equations

$$P_{sc} = |V_{sc}|^2 G_{sc} - |V_{sc}||V_j|\left[G_{sc}\cos(\delta_{sc} - \delta_j) + B_{sc}\sin(\delta_{sc} - \delta_j)\right] \quad (7.123)$$

$$Q_{sc} = -|V_{sc}|^2 B_{sc} - |V_{sc}||V_j|\left[G_{sc}\sin(\delta_{sc} - \delta_j) - B_{sc}\cos(\delta_{sc} - \delta_j)\right] \quad (7.124)$$

To simplify these equations, let us assume that the STATCOM is lossless (thus $G_{sc} = 0$) and there is no capability of the STATCOM for active power flow (thus

$P_{sc} = 0$). Also $\delta_{sc} \cong \delta_j$.

$$\therefore \qquad Q_{sc} = -|V_{sc}|^2 B_{sc} - |V_{sc}||V_j| B_{sc} \qquad (7.125)$$

The power mismatch equation can be written now as

$$\begin{bmatrix} \Delta P_j \\ \Delta Q_{sc} \end{bmatrix} = \begin{bmatrix} \dfrac{\partial P_j}{\partial \delta_j} & \dfrac{\partial P_j}{\partial |V_{sc}|} \\ \dfrac{\partial Q_{sc}}{\partial \delta_j} & \dfrac{\partial Q_{sc}}{\partial |V_{sc}|} \end{bmatrix} \begin{bmatrix} \Delta \delta_j \\ \Delta |V_{sc}| \end{bmatrix} \qquad (7.126)$$

At the end of iteration p, the variable voltage $|V_{sc}|$ can be corrected as

$$|V_{sc}|^{(p+1)} = |V_{sc}|^{(p)} + \Delta |V_{sc}|^{(p)} \qquad (7.127)$$

7.17 SERIES–SERIES CONTROLLERS

Any Standard Series Controller (FACTS device) may be suitably connected with another type of series FACTS controller to form a series-series controller (Figure 7.36). As a typical example we may think of a Thyristor Controlled Series Capacitor (TCSC) in series with a Thyristor Switched Series Capacitor (TSSC). It is reasonable to arrange this series connection such that one module has smooth thyristor control while the other could be thyristor switched control. Though not very common in use, series-series controller (Figure 7.36) may be applied for the control of power in double circuit ac lines.

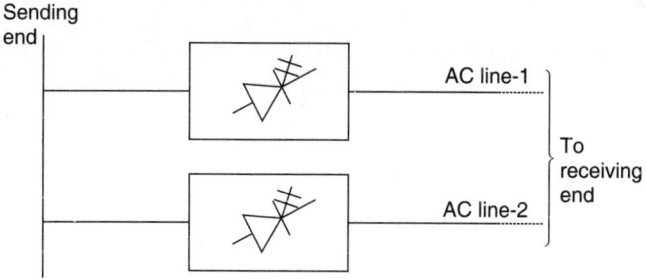

Figure 7.36 Schematic of a series–series controller.

7.18 COMBINED SHUNT–SERIES CONNECTED CONTROLLERS

7.18.1 Static Phase Shifter (SPS)

A *Static Phase Shifter* (SPS) or *Thyristor Controlled Phase Shifting Transformer* (TC PST) is basically a phase shifting transformer adjusted by thyristor switches to provide smooth variation in phase angle. Figure 7.37 shows such an arrangement where phase shifting is arranged by adding a quadrature voltage vector in series with the phase voltage. This quadrature voltage vector is derived from the other two phases via shunt-connected transformers. The prependicular series voltage is made variable using the thyristor-based voltage controller. SPS is connected in all the three phases. Using proper control circuitry, it is possible to even reverse the voltage so that phase shift in either direction is possible.

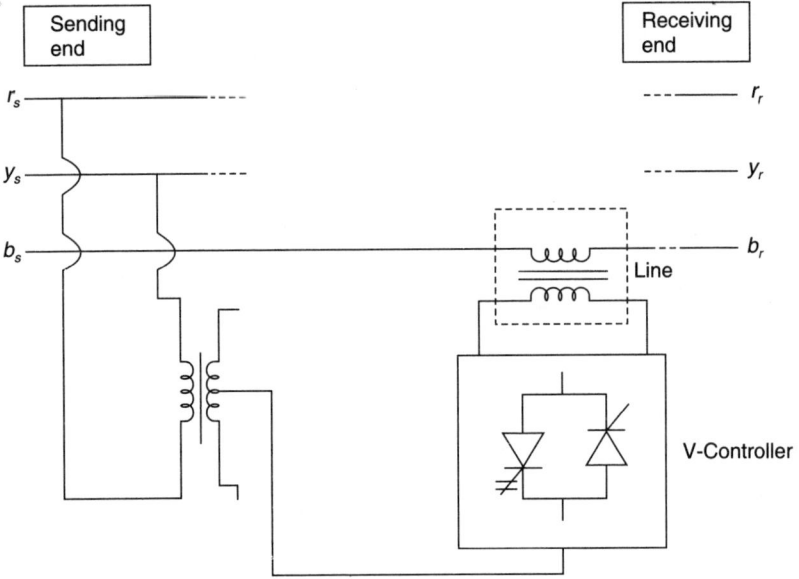

Figure 7.37 Schematc of an SPS controller.

Let us denote the sending end as suffix s and the receiving-end as suffix r. We assume a lossless system. The equivalent circuit of the SPS is shown in Figure 7.38.

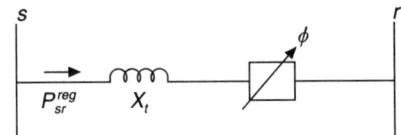

Figure 7.38 Equivalent circuit of representation of SPS.

Let us assume I_s and I_r to be the sending-end (s) and receiving-end (r) currents with V_s and V_r as the respective voltages, and X_t the series transformer reactance. We can express the tap-changing variables n_V and n_I as

$$n_V = 1 \angle \phi = \cos \phi + j \sin \phi \tag{7.128}$$

$$n_1 = 1 \angle -\phi = \cos \phi - j \sin \phi \tag{7.129}$$

(n_V and n_1 are related to each other by complex conjugate relation)

Also,
$$\begin{bmatrix} I_s \\ I_r \end{bmatrix} = \frac{1}{X_t} \cdot \begin{bmatrix} 1 & -(\cos\phi + j\sin\phi) \\ -(\cos\phi - j\sin\phi) & 1 \end{bmatrix} \begin{bmatrix} V_s \\ V_r \end{bmatrix} \quad (7.130)$$

The active and reactive powers injected at sending end (s) and receiving-end (r) buses of the SPS are obtained as

$$P_s = |V_s|^2 G_{ss} + |V_s||V_r|[G_{sr} \cos(\delta_s - \delta_r) + B_{sr} \sin(\delta_s - \delta_r)] \quad (7.131)$$

$$Q_s = -|V_s|^2 B_{ss} - |V_s||V_r|[B_{sr} \cos(\delta_s - \delta_r) - G_{sr} \sin(\delta_s - \delta_r)] \quad (7.132)$$

Aslo, $\quad P_r = |V_r|^2 G_{rr} + |V_r||V_s|[G_{rs} \cos(\delta_r - \delta_s) + B_{rs} \sin(\delta_r - \delta_s)] \quad (7.133)$

$$Q_r = -|V_r|^2 B_{rr} - |V_r||V_s|[B_{rs} \cos(\delta_r - \delta_s) - G_{rs} \sin(\delta_r - \delta_s)] \quad (7.134)$$

The power mismatch equations are obtained as

$$\begin{bmatrix} \Delta P_s \\ \Delta Q_s \\ \Delta P_r \\ \Delta Q_r \\ \Delta P_{sr} \end{bmatrix} = \begin{bmatrix} \frac{\partial P_s}{\partial \delta_s} & \frac{\partial P_s}{\partial |V_s|} & \frac{\partial P_s}{\partial \delta_r} & \frac{\partial P_s}{\partial |V_r|} & \frac{\partial P_s}{\partial \phi} \\ \frac{\partial Q_s}{\partial \delta_s} & \frac{\partial Q_s}{\partial |V_s|} & \frac{\partial Q_s}{\partial \delta_r} & \frac{\partial Q_s}{\partial |V_r|} & \frac{\partial Q_s}{\partial \phi} \\ \frac{\partial P_r}{\partial \delta_s} & \frac{\partial P_r}{\partial |V_s|} & \frac{\partial P_r}{\partial \delta_r} & \frac{\partial P_r}{\partial |V_r|} & \frac{\partial P_r}{\partial \phi} \\ \frac{\partial Q_r}{\partial \delta_s} & \frac{\partial Q_r}{\partial |V_s|} & \frac{\partial Q_r}{\partial \delta_r} & \frac{\partial Q_r}{\partial |V_r|} & \frac{\partial Q_r}{\partial \phi} \\ \frac{\partial P_{sr}}{\partial \delta_s} & \frac{\partial P_{sr}}{\partial |V_s|} & \frac{\partial P_{sr}}{\partial \delta_r} & \frac{\partial P_{sr}}{\partial |V_r|} & \frac{\partial P_{sr}}{\partial \phi} \end{bmatrix} \begin{bmatrix} \Delta \delta_s \\ \Delta |V_s| \\ \Delta \delta_r \\ \Delta |V_r| \\ \Delta \delta \end{bmatrix}$$

(7.135)

Here,
δ_s = angle of V_s at sending end
δ_r = angle of V_r at receiving-end
ϕ = phase shifter angle (variable).

δ includes the phase shifter angle (ϕ) and it has been assumed that the SPS has phase angle tapping in the primary winding. The phase angle tapping can also be assumed in secondary winding but in that case the Jacobian terms are to be derived with respect to δ_r.

Also, $\qquad P_{sr}^{cal} = \dfrac{|V_s||V_r|}{X_t} \sin(\delta_s - \delta_r - \phi) \qquad (7.136)$

Note that P_{sr} is variable / controllable, and $\Delta P_{sr} = P_{sr}^{reg} - P_{sr}^{cal}$ and P_{sr}^{cal} is obtained from Eq. (7.136). P_{sr}^{reg} is the desired real power transfer value. Thus the Jacobian can be deriverd. The variable δ can be updated as follows:

$$\delta^{(p+1)} = \delta^{(p)} + \Delta\delta^{(p)} \tag{7.137}$$

The phasor diagram of the system is shown in Figure 7.39.

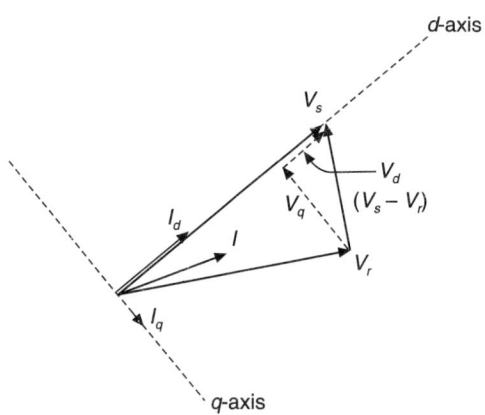

Figure 7.39 Phasor diagram of SPS operation.

The power transmission results in a difference in magnitude between the terminal voltages V_s and V_r and also a shift in phase angle. Phasor voltage difference $(V_s - V_r)$ appears across X_t resulting in the current through the transmission line. Phasor $(V_s - V_r)$ consists of the inductive drop and can be decomposed into two components, one in phase and the other in quadrature with V_s. These voltage components determine the reactive and real powers supplied by the sending-end side. It is possible to control both the magnitude and the phase of $(V_s - V_r)$, thus controlling both $|V_d|$ and $|V_q|$. Both the real and reactive power flows are thus controllable using SPS.

7.18.2 Unified Power Flow Controller (UPFC)

UPFC is one of the most advanced FACTS devices and is a combination of static synchronous compensator (STATCOM) and a static series synchronous compensator (SSSC). The two devices are coupled through a dc link and the combination allows bidirectional flow of real power between the series output of SSSC and the shunt output of STATCOM. This controller (UPFC) has the facility to provide concurrent real and reactive series line compensation without any external electric energy source. UPFC can have angularly controlled series voltage injection for transmission voltage control in addition to the line impedance and power angle control. Thus UPFC is able to control both the real and the reactive power flows in lines and may also be used as independently controlled shunt reactive compensation.

Figure 7.41 represents the schematic of a UPFC which contains a STATCOM with a SSSC. The active power flow for the series unit (SSSC) is obtained from the line itself through the shunt unit (STATCOM). STATCOM is employed for voltage (or reactive power) control while SSSC is utilized for real power control. UPFC is a complete FACTS controller for both real and reactive power flow control in a line.

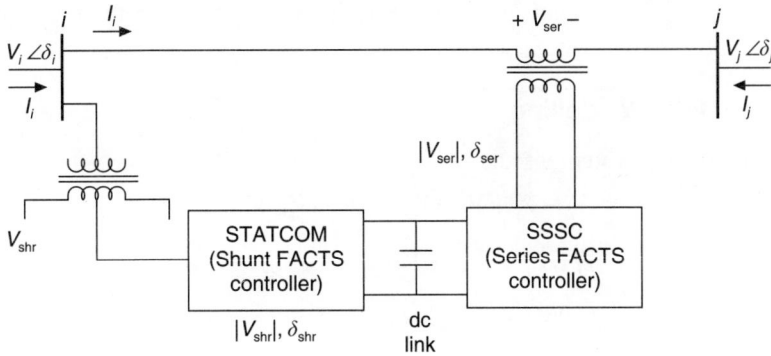

Figure 7.40 Schematic of a UPFC.

An additional storage device (viz. a superconducting magnet connected to the dc link) through an electronic interface would provide the enhancement in the capability of UPFC in real power flow control.

The active power required for the series-converter is drawn by the shunt-converter from the ac bus (i) and supplied by the dc link. The inverted ac voltage (V_{ser}) at the output of the series converter is added to the sending-end voltage V_i at line side to boost the voltage at the jth bus. It may be noted here that $|V_{ser}|$ provides regulation while δ_{ser} determines the mode of power flow control.

The following conditions are important in understanding the operation of UPFC with reference to its equivalent circuit (Figure 7.41).

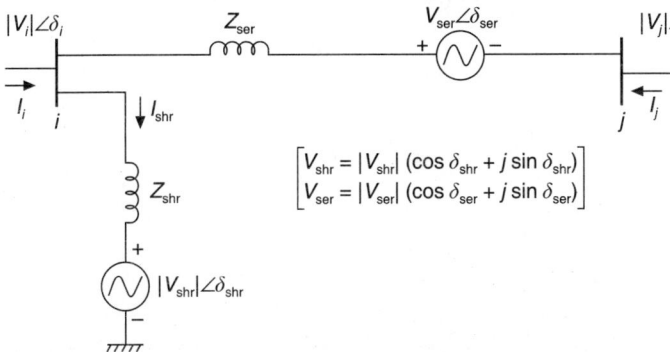

Figure 7.41 Equivalent circuit of a UPFC between two buses i and j.

- If δ_{ser} is in phase with δ_i, no active power flow takes place between the i and jth buses. Reactive power flow can be controlled by varying $|V_{ser}|$.
- If δ_{ser} is in quadrature with δ_i, active power flow can be controlled between the ith and jth buses by controlling δ_{ser}. No reactive power flow will occur between ith and jth buses.
- If δ_{ser} is in quadrature with line current angle, then it can also control the active power flow.
- If δ_{ser} is in between 0° and 90°, it can control both the real and the reactive power flows in the line.

7.18.3 UPFC Modelling

Based on the equivalent circuit as shown in Figure 7.42, we have

$$I_i = (V_i - V_j - V_{ser}) Y_{ser} + (V_i - V_{shr}) Y_{shr}$$

$$= V_i (Y_{ser} + Y_{shr}) - V_j Y_{ser} - V_{ser} Y_{ser} - V_{shr} Y_{shr}$$

and

$$I_j = (-V_i + V_j + V_{ser}) Y_{ser}$$

i.e.

$$\begin{bmatrix} I_i \\ I_j \end{bmatrix} = \begin{bmatrix} (Y_{ser} + Y_{shr}) & -Y_{ser} & -Y_{ser} & -Y_{shr} \\ -Y_{ser} & Y_{ser} & Y_{ser} & 0 \end{bmatrix} \begin{bmatrix} V_i \\ V_j \\ V_{ser} \\ V_{shr} \end{bmatrix}$$

Also,

$$\begin{bmatrix} S_i \\ S_j \end{bmatrix} = \begin{bmatrix} V_i & 0 \\ 0 & V_j \end{bmatrix} \begin{bmatrix} I_i^* \\ I_j^* \end{bmatrix}$$

$$= \begin{bmatrix} V_i & 0 \\ 0 & V_j \end{bmatrix} \begin{bmatrix} (Y_{ser} + Y_{shr})^* & -Y_{ser}^* & -Y_{ser}^* & -Y_{shr}^* \\ -Y_{ser}^* & Y_{ser}^* & Y_{ser}^* & 0 \end{bmatrix} \begin{bmatrix} V_i^* \\ V_j^* \\ V_{ser}^* \\ V_{shr}^* \end{bmatrix}$$

$$= \begin{bmatrix} V_i & 0 \\ 0 & V_j \end{bmatrix}$$

$$\begin{bmatrix} (G_{ii} - jB_{ii}) & (G_{ij} - jB_{ij}) & (G_{ij} - jB_{ij}) & (G_{i0} - jB_{i0}) \\ (G_{ji} - jB_{ji}) & (G_{jj} - jB_{jj}) & (G_{jj} - jB_{jj}) & 0 \end{bmatrix} \begin{bmatrix} V_i^* \\ V_j^* \\ V_{ser}^* \\ V_{shr}^* \end{bmatrix}$$

or $\begin{bmatrix} P_i + jQ_i \\ P_j + jQ_j \end{bmatrix}$

$$= \begin{bmatrix} V_i & 0 \\ 0 & V_j \end{bmatrix} \begin{bmatrix} (G_{ii} - jB_{ii}) & (G_{ij} - jB_{ij}) & (G_{ij} - jB_{ij}) & (G_{i0} - jB_{i0}) \\ (G_{ji} - jB_{ji}) & (G_{jj} - jB_{jj}) & (G_{jj} - jB_{jj}) & 0 \end{bmatrix} \begin{bmatrix} V_i^* \\ V_j^* \\ V_{ser}^* \\ V_{shr}^* \end{bmatrix}$$

(7.138)

$\therefore \quad P_i = |V_i|^2 G_{ii} + |V_i||V_j| \{ G_{ij} \cos(\delta_i - \delta_j) + B_{ij} \sin(\delta_i - \delta_j) \}$

$\quad + |V_i||V_{ser}| \{ G_{ij} \cos(\delta_i - \delta_{ser}) + B_{ij} \sin(\delta_i - \delta_{ser}) \}$

$\quad + |V_i||V_{shr}| \{ G_{i0} \cos(\delta_i - \delta_{shr}) + B_{i0} \sin(\delta_i - \delta_{shr}) \}$ (7.139)

$Q_i = -|V_i|^2 B_{ii} + |V_i||V_j| \{ G_{ij} \sin(\delta_i - \delta_j) - B_{ij} \cos(\delta_i - \delta_j) \}$

$\quad + |V_i||V_{ser}| \{ G_{ij} \sin(\delta_i - \delta_{ser}) - B_{ij} \cos(\delta_i - \delta_{ser}) \}$

$\quad + |V_i||V_{shr}| \{ G_{i0} \sin(\delta_i - \delta_{shr}) - B_{i0} \cos(\delta_i - \delta_{shr}) \}$ (7.140)

$P_j = |V_j|^2 G_{jj} + |V_j||V_i| \{ G_{ji} \cos(\delta_j - \delta_i) + B_{ji} \sin(\delta_j - \delta_i) \}$

$\quad + |V_j||V_{ser}| \{ G_{jj} \cos(\delta_j - \delta_{ser}) + B_{jj} \sin(\delta_j - \delta_{ser}) \}$ (7.141)

$Q_j = -|V_j|^2 B_{jj} + |V_j||V_i| \{ G_{ji} \sin(\delta_j - \delta_i) - B_{ji} \cos(\delta_j - \delta_i) \}$

$\quad + |V_j||V_{ser}| \{ G_{jj} \sin(\delta_j - \delta_{ser}) - B_{jj} \cos(\delta_j - \delta_{ser}) \}$ (7.142)

The active and reactive powers of the series converter (SSSC) are as follows:

$S_{ser} = P_{ser} + jQ_{ser} = V_{ser} I_j^* = V_{ser} \left[Y_{ji}^* V_i^* + Y_{jj}^* V_j^* + Y_{jj}^* V_{ser}^* \right]$ (7.143)

$\therefore \quad P_{ser} = |V_{ser}|^2 G_{jj} + |V_{ser}||V_i| \{ G_{ji} \cos(\delta_{ser} - \delta_i) + B_{ji} \sin(\delta_{ser} - \delta_i) \}$

$\quad + |V_{ser}||V_j| \{ G_{jj} \cos(\delta_{ser} - \delta_j) + B_{jj} \sin(\delta_{ser} - \delta_j) \}$ (7.143a)

$Q_{ser} = -|V_{ser}|^2 B_{jj} + |V_{ser}||V_i| \{ G_{ji} \sin(\delta_{ser} - \delta_i) - B_{ji} \cos(\delta_{ser} - \delta_i) \}$

$\quad + |V_{ser}||V_j| \{ G_{jj} \sin(\delta_{ser} - \delta_j) - B_{jj} \cos(\delta_{ser} - \delta_j) \}$ (7.143b)

The active and reactive powers of the shunt converter (STATCOM) are obtained as

$S_{shr} = P_{shr} + jQ_{shr} = V_{shr} I_{shr}^* = -V_{shr} Y_{shr}^* \left(V_{shr}^* - V_i^* \right)$

$$\therefore P_{shr} = -|V_{shr}|^2 G_{i0} + |V_{shr}||V_i|\{G_{i0}\cos(\delta_{shr}-\delta_i)+B_{i0}\sin(\delta_{shr}-\delta_i)\} \quad (7.144)$$

$$Q_{shr} = |V_{shr}|^2 B_{i0} + |V_{shr}||V_i|\{G_{i0}\sin(\delta_{shr}-\delta_i)-B_{i0}\cos(\delta_{shr}-\delta_i)\} \quad (7.145)$$

Since we assume lossless converters, the UPFC neither absorbs nor injects active power with respect to the AC system. Hence the constraint equation is

$$P_{shr} + P_{ser} = 0 \quad (7.146)$$

In order to get the linearised model of the system using the power mismatch from, let we assume UPFC is connected to node i and the power system is connected to node j. UPFC is required to control the voltage of node i and the active power flows from node j to node i. Reactive power is injected at node j. Here, we can write

$$\begin{bmatrix} \Delta P_i \\ \Delta P_j \\ \Delta Q_i \\ \Delta Q_j \\ \Delta P_{ji} \\ \Delta Q_{ji} \\ \Delta P \end{bmatrix} = \begin{bmatrix} \frac{\partial P_i}{\partial \delta_i} & \frac{\partial P_i}{\partial \delta_j} & \frac{\partial P_i}{\partial |V_{shr}|} & \frac{\partial P_i}{\partial |V_j|} & \frac{\partial P_i}{\partial \delta_{ser}} & \frac{\partial P_i}{\partial |V_{ser}|} & \frac{\partial P_i}{\partial \delta_{shr}} \\ \frac{\partial P_j}{\partial \delta_i} & \frac{\partial P_j}{\partial \delta_j} & 0 & \frac{\partial P_j}{\partial |V_j|} & \frac{\partial P_j}{\partial \delta_{ser}} & \frac{\partial P_j}{\partial |V_{ser}|} & 0 \\ \frac{\partial Q_i}{\partial \delta_i} & \frac{\partial Q_i}{\partial \delta_j} & \frac{\partial Q_i}{\partial |V_{shr}|} & \frac{\partial Q_i}{\partial |V_j|} & \frac{\partial Q_i}{\partial \delta_{ser}} & \frac{\partial Q_i}{\partial |V_{ser}|} & \frac{\partial Q_i}{\partial \delta_{shr}} \\ \frac{\partial Q_j}{\partial \delta_i} & \frac{\partial Q_j}{\partial \delta_j} & 0 & \frac{\partial Q_j}{\partial |V_j|} & \frac{\partial Q_j}{\partial \delta_{ser}} & \frac{\partial Q_j}{\partial |V_{ser}|} & 0 \\ \frac{\partial P_{ji}}{\partial \delta_i} & \frac{\partial P_{ji}}{\partial \delta_j} & 0 & \frac{\partial P_{ji}}{\partial |V_j|} & \frac{\partial P_{ji}}{\partial \delta_{ser}} & \frac{\partial P_{ji}}{\partial |V_{ser}|} & 0 \\ \frac{\partial Q_{ji}}{\partial \delta_i} & \frac{\partial Q_{ji}}{\partial \delta_j} & 0 & \frac{\partial Q_{ji}}{\partial |V_j|} & \frac{\partial Q_{ji}}{\partial \delta_{ser}} & \frac{\partial Q_{ji}}{\partial |V_{ser}|} & 0 \\ \frac{\partial P}{\partial \delta_i} & \frac{\partial P}{\partial \delta_j} & \frac{\partial P}{\partial |V_{shr}|} & \frac{\partial P}{\partial |V_j|} & \frac{\partial P}{\partial \delta_{ser}} & \frac{\partial P}{\partial |V_{ser}|} & \frac{\partial P}{\partial \delta_{shr}} \end{bmatrix} \begin{bmatrix} \Delta \delta_i \\ \Delta \delta_j \\ \Delta |V_{shr}| \\ \Delta |V_j| \\ \Delta \delta_{ser} \\ \Delta |V_{ser}| \\ \Delta \delta_{shr} \end{bmatrix}$$

(7.147)

It has been assumed that node j is the PQ node while ΔP is the power mismatch given by the constrained equation.

An extensive algorithm is needed for the solution of power flow equation using UPFC. Good starting conditions for all the UPFC state variables are also an important requirements to ensure convergence.

7.18.4 Interphase Power Controller (IPC)

IPC is another type of combined shunt and series controller. It is a series-connected FACTS device having capability of controlling both the active and reactive power in each phase. The active and reactive power can be set independently by adjusting the phase shift and/or the branch impedance.

Researches in FACTS devices reveal that a combination of series and shunt controllers may be a very effective tool in both real and reactive power control. However, from applicability point of view, SVC is a very widely accepted FACTS device followed by STATCOM and TCSC. Research is going on across the globe with FACTS controllers and in most of the research findings we observe that FACTS devices are very much suitable for enhancing complete stability, reliability and controllability of complex power systems.

7.19 ADVANTAGES OF FACTS DEVICES

We can now generalise the advantages of FACTS devices as follows:
- Control of power flow (both active and reactive), as desired and within limits, is possible.
- Reduction in voltage drop in power lines is possible. Regulation can be improved.
- Reduction in reactive burden on line allows more flow of active power in lines.
- Loadability of lines is increased.
- Voltage stability and voltage security are enhanced.
- Security of tie lines connecting two sub-grids is increased.
- Transient stability is increased.
- Short-circuit currents and overloads can be controlled up to a certain limit.
- Generation cost reduces.
- Passive compensation requirement reduces.

EXERCISES

1. What is the role of a OLTC transformer in voltage control of a transmission line?
2. How does the performance of a tie transformer very while operating between two power systems of varying short-circuit strength?
3. Derive expressions to show the limited voltage control capability of a transformer.
4. Discuss the role of a OLTC transformer on voltage stability.
5. What are the methods of improving voltage stability using passive compensation?
6. How does a fixed shunt capacitor improve the voltage stability of a load bus?
7. What is passive series compensation? Determine the criterion of line loadability limit for series compensated lines.
8. What is the reactive power requirement at the load-end of a line when the angular separation reaches the limiting value? How do you find the critical

angular separation with finite VAR reserve for voltage controlled and for voltage uncontrolled load bus for series compensated lines?

9. How do you justify the use of capacitive series compensation on voltage instability of EHV long lines?
10. Discuss the effect of location of series capacitor on voltage stability.
11. What are the effects of capacitor bussing arrangement and contingency on voltage stability of EHV series compensated lines?
12. Develop expressions to find the optimal amount of load shedding to have voltage stability following a contingency.
13. Describe the model of a synchronous condenser at the load bus with limited control of voltage at the receiving end.
14. Classify the different types of *FACTS* controllers.
15. Describe the models of shunt connected FACTS controllers. Why is STATCOM superior to SVC?
16. What is TCSC? How do you model it? What are the other series connected FACTS controllers?

References

Abbey, C. and G. Joos, Effect of low voltage ride through (LVRT) characteristic on voltage stability, Power Engineering Society General Meeting, *2005 IEEE*.

Ambriz-Perez, H., E. Ache, and C.R. Fuerte-Esquivel, Advanced SVC model for N-R load flow and newton optimal power flow studies—*IEEE Transactions on Power System*, Vol. 15, No. 1, February 2000.

Cai, Li-Jun and I. Erlich, Power system static voltage stability analysis considering all active and reactive power controls—Singular value approach, Power Tech., *2007 IEEE Lausanne*, pp. 367–373, Publication Date: 1–5, July 2007.

Chakrabarti, A. and A.K. Mukhopadhyay, An optimal load shedding scheme to maintain voltage stability following a line contingency in a longitudinal power supply system, *Journal Modelling Simulation and Control*, A. AMSE, France, Vol. 42, No. 21, 1992.

Chakrabarti, A. and A.K. Mukhopadhyay, Contingency simulation on the operation of series capacitor compensated EHV lines at voltage stable state, *Proc. of 29th SICE-IEEE-KAAC*, International Conference, Tokyo, Japan, 1990.

Chakrabarti, A. and A.K. Mukhopadhyay, Load end real and reactive power contour and shunt capacitive support at voltage stability limit of an EHV power system, *Proc. of 28th SICE-IEEE-KAAC*, Annual International Conference, Japan, 1989.

Chakrabarti, A. and A.K. Mukhopadhyay, Voltage stability of longitudinal power systems, *Journal of Electrical Engineering*, Institution of Engineers (I), Pt EL-3, 1990.

Chakrabarti, A. and P. Sen, Application of computers and microprocessors in power system control and protection, *Proc. VIII NPSC*, New Delhi, India, 1994.

Chakrabarti, A. and S. Haldar, *Power System Analysis Operation and Control*, PHI Learning, New Delhi, 2010.

Chanda, C.K., S. Dey, A. Chakrabarti, and A.K. Mukhopadhyay, Determination of bus Security governed by sensitivity indicator in a reactive power constraint longitudinal power supply (LPS) system, *Indian Journal of Engineering and Materials Sciences*, Vol. 9, August 2002, pp. 260–264.

Chibbo, A.M. and M.R. Irving, voltage collapse proximity indicator: Behaviors and implication, *IEE Proc.*, May 1992.

Chow, J.C., R. Fischl, and H. Yan, On the evaluation of voltage collapse criteria, *IEEE Transaction on Power System*, Vol. 5, No. 2, May 1990.

Cutsen, T. Van and C. Vournas, *Voltage Stability of Electric Power System*, Kluwer Academic Publisher, Boston, 2001.

Davidson, D.R., Long term dynamic response of power system and analysis of major disturbances, *IEEE Transaction on PAS*, Vol. PAS-94, No. 3, 1975.

Edris, A.A., Controllable VAR compensation: A potential solution to loadability problems of low capacity power systems, *IEEE Transaction on Power System*, Vol. PWRS-2, No. 3, 1987.

El-Sadek, M.Z., M. Dessouky, G.A. Mahmoud, and W.I. Rashed, Series capacitors combined with static VAR compensators for enhancement of steady-state voltage stabilities, Electrical Power Systems Research, 44 (1998), pp. 137–143.

Flourentzou, N. and G.D. Demetriades, VSC—Based HVDC power transmission system—An overview, *IEEE Transaction on Power Electronics*, Vol. 24, Issue 3, March 2009, pp. 592–602.

Godari, T.F. and H.G. Puttgen, A reactive path concept applied within a voltage control expert system, *IEEE Transaction on Power System*, Vol. 6, No. 2, February 1991.

Hammad, A.E. et al., Prevention of transient voltage instabilities due to induction motor loads by static VAR compensations, *IEEE Transaction on Power System*, Vol. PWRS-1, No. 3, August 1989.

Haque, M.H., Determination of steady state voltage stability limit of a power system in the presence of SVC, *Power Tech Proceedings, 2001, IEEE Porto*, Vol. 2.

Haque, M.H., On-line monitoring of maximum permissible loading of a power system within voltage stability limits, *IEE Proc.*, Gener. Transm. Distrib., January 2003, Vol. 150, Issue 1, pp. 107–112.

Hill, D.J., Nonlinear dynamic load models with recovery for voltage stability studies, *IEEE Transaction on Power System*, Vol. 8, No. 1, February 1994.

Hingorani, N.G. and L. Gyugyi, *Understanding FACTS—Concept and Technology of Flexible AC Transmission Systems*, IEEE Press, 1999.

Iliceto, F. and E. Cinieri, Comperative analysis of series and shunt compensation scheme for AC transmission system, *IEEE Transaction*, PAS Vol. 6, No. 96, 1977.

Indulkar, C.S. and B. Viswanathan, Series compensation in EHV transmission lines, *Journal of Institute of Engineers*, Vol. 65, Journal EL-2 and 3, 1984.

Kataoka, Y., M. Watanabe, and S. Iwamoto, A new voltage stability index considering voltage limits, Power Systems Conference and Exposition, *PSCE'06, IEEE*, October 29–November 1 2006.

Kothari, D.P. and I.J. Nagrath, *Modern Power System Analysis*, 3rd ed., Tata McGraw Hill, New Delhi, 2003.

Kumari, M. Sailaja and M. Sydulu, A novel load flow approach for voltage stability index calculation and adjustment of static VAR compensator parameters, *Power India Conference, 2006 IEEE*.

Lachs, W.R., System reactive power limitation, Paper A 79-015-9, *IEEE 1979 Winter Power Conference*, USA.

Miller, T.J.E., *Reactive Power Control in Electric Systems*, John-Wiley & Sons, 1982.

Milosevic, B. and M. Begovic, Voltage-stability protection and control using a wide-area network of phasor Measurements, *IEEE Transaction on Power System*, February 2003, Vol. 18, Issue 1.

Moghavvemi, M. and F.M. Omar, Real time contingency evaluation and ranking technique, *IEE Proc. Generator, Transmission and Distribution*, Vol. 145, No. 5, September 1998.

Moghavvemi, M. and F.M. Omar, Technique for contingency monitoring and voltage collapse prediction, *IEE Proc. Generator, Transmission and Distribution*, Vol. 145, No. 6, November 1998.

Moghavvemi, M. and M.O. Faruque, Effects of FACTS devices on static voltage stability, *TENCON 2000 Proceedings*, Vol. 2, pp. 356–362.

Mustafa, M.W. and A.F. Abdul Kadir, A modified approach for load flow analysis of Integrated AC-DC power systems, *IEEE 2000*, Vol. II, pp. 108–113.

Nagrath, I.J. and D.P. Kothari, *Power System Engineering*, Tata McGraw-Hill, New Delhi, 1994.

Opoku, G., Optimal power system VAR planning, *IEEE Transaction on Power System*, Vol. 5, No. 1, February 1990.

Reis, C. and F.P.M. Barbosa, A comparison of voltage stability indices, Electrotechnical Conference, *MELECON IEEE*, May 2006.

Sanghavi, H.A. and S.K. Banerjee, Load flow analysis of an integrated AC–DC power network, *IEEE* 1989, pp. 746–751.

Sauer, P. and M.A. Pai, *Power System Dynamics and Stability*, Pearson Education, Asia, 2002.

Savulesca, S.C., Quantitative indices for the system voltage reactive power control, *IEEE Transaction on PAS*, Vol. PAS-95 July/Aug 1976.

Sekine, Y. et al., Cascaded voltage collapse, *IEEE Transaction on Power System*, Vol. 5, No. 1, February 1990.

Semlyn, A., B. Gao, and W. Janischewskyj, Calculation of extreme loading condition of a power system for the assessment of voltage stability, *IEEE Transaction on Power System*, Vol. 6, No. 2, February 1991.

Sinha. A.K. and D. Hazarika, A comparative study of voltage stability indices in a power system, Electrical Power and Energy Systems, 22 (2000), pp. 589–596.

Smolinski, W.J., Equivalent circuit analysis of power system reactive power and voltage control problems, *IEEE Transaction on PAS*, Vol. PAS-100, No. 2, Feb. 1981.

Stott, B., Review of load flow calculation methods—*Proceedings of the IEEE*, Vol. 62, No. 7, July 1974.

Taylor, C.W., *Power System Voltage Stability (EPRS)*, Tata McGraw-Hill Inc. USA, 1994.

Tiranuchit, A. and R.J. Thomas, A posturing strategy against voltage instability in electric power system, *IEEE Transaction on Power System*, Vol. PWRS-3, No. 1, 1988.

VanCutsem, T., A method to compute reactive power margins with respect to voltage collapse, *IEEE Transaction on Power System*, Vol. 6, No. 2, February 1991.

Vournas, C. and M. Karystianos, Load tap changers in emergency and preventive voltage stability control, *IEEE Transaction on Power System*, February 2004, Vol. 19, Issue 2.

Wang, Lei, Meir Kiein, Solomon Yirga, and Prabha Kundur, Dynamic reduction of large power system for stability studies, *IEEE Transaction on Power System*, Vol. 12, No. 2, May 1997, pp. 889–895.

Wen, J.Y., Q.H. Wu, D.R. Turner, S.J. Cheng, and J. Fitch, Optimal coordinated voltage control for power system voltage stability, *IEEE Transaction on Power System*, May 2004, Vol. 19, Issue 2.

Yao, Liangzhong, P. Cartwright, L. Schmitt, and Xiao-Ping Zhang, Congestion Management of Transmission Systems Using FACTS, *2005 IEEE/PES Transmission and Distribution Conference and Exhibition*, Asia and Pacific Dalian, China.

Yome, A., N. Mithulananthan, and Kwang Y. Le, A maximum loading margin method for static voltage stability in power systems, *IEEE Transaction on Power System*, Vol. 21, No. 2, February 2006.

Zarate, L.A.L., C.A. Castro, J.L.M. Ramos, and E.R. Ramos, Fast computation of voltage stability security margins using nonlinear programming techniques, *IEEE Transaction on Power System*, Vol. 21, February 2006.

Index

Asynchronous loads, 32
Attenuation constant, 4
AVR control, 86

Battery Energy Storage System (BESS), 239
Bus sensitivity, 172

Characteristic impedance, 4
Compensation devices, 86
Complex power, 6
Contingency ranking, 179
    1P1Q method, 177
Continuation load-flow mehtod, 159
    prediction-correction, 159
Critical bus, 174
Critical clearing angle, 99
Critical receiving-end voltage, 166

Distributed parameters, 4
Dynamic stability, 52
Dynamic voltage stability, 91

Electrical length, 5
Exponent load model, 81

FACTS, 232
    Thyristor Controlled Series Capacitor (TCSC), 235
Fast Decoupled Load Flow (FDLF) algorithm, 40
Fast voltage security index, 131
Ferranti effect, 21

Fixed Capacitor Thyristor Controlled Reactor (FCTCR), 239
Flat voltage profile, 24

Half line, 22
Hessian matrix, 172
Hopf bifurcation, 77
HVDC, 71

Interphase Power Controller (IPC), 250

Jacobian, 101
Jacobian matrix, 77, 172

Lagrange multiplier functions, 169
Line loadability, 35
    nodal indicator of robustness, 35
    nodal short-circuit strengths, 28
Line loadability limit, 210
Line outage factor, 131
Line quality factor, 131
Loadability, 17, 203
Load coefficients, 213
Load shedding, 223
Load voltage indicator, 123
Long line, 4
Longitudinal Power System (LPS), 90
Lossless line, 4

Medium lines, 2
Minimum singular value method, 156
Modal analysis, 162

258   Index

Natural impedance, 16

On load tap changer, 83, 143
    restoration of load, 85
Optimal Power Flow (OPF), 167
Optimisation method, 158

Performance index, 178, 224
Phase constant, 4
Point of collapse method, 157
    saddle node bifurcation, 158
Polynomial model, 81
Power circle diagram, 11
Power system security, 148
    emergency, 149
    extreme emergency, 149
    normal alert, 149
    restorative, 149
Power triangle, 6
Primary constants, 1
Propagation constant, 4
PV-curves, 74

Reactive power, 6
Reactive power deficit, 56
Reactive power limits, 78
Reactive power sensitivity, 43
    degree of stability, 43
Reactive power stability, steady state, 47
Reactive power surplus, 56
Reactive power transient stability, 51
Reactive power-constrained line, 102
Real power, 6
Real power steady state stability margin, 67
Real power stability, steady state, 46
Real power transient stability, 50
Rotor angle, 66

Security assessment, 148, 151
Security boundary, 153
Sensitivity, 43
Sensitivity coefficients, 128
Series compensation, 202
Series controllers, 233
Series reactive loss, 43

Series-series controllers, 233
Series-shunt controllers, 233
Short-circuit capacity, 35, 120
Short-circuit current, 54
Short-circuit driving point, 7
Short-circuit strength, 54
Short line, 2
Shunt capacitor compensation, 197
Shunt controllers, 233
Small-signal stability, 71
Soft constraints, 169
Static Synchronous Series Compensator (SSSC), 234
Stability, 66
Stability margin, 127
Static contingency, 178
Static Phase Shifter (SPS), 243
Static Synchronous Compensator (STATCOM), 241
Static Synchronous Generator (SSG), 239
Static VAR Compensator (SVC) 71, 238
Strategic load shedding, 60
Superconducting magnetic energy storage system, 239
Supervisory control and data acquisition, 151
Surge impedance loading, 15
Synchronous condenser, 228
Synchronous generator, 78
    active power limits, 78
Synchronous loads, 32
Synchronous tie, 22

Thyristor Controlled Phase Shifting Transformer (TC PST), 243
Thyristor Controlled Series Capacitor, 235
Thyristor Controlled Series Reactor (TCSR), 237
Thyristor Switched Capacitor Thyristor Controller Reactor (TSC TCR), 239
Thyristor switched series reactor, 238
Transfer admittance, 7
Transfer capacity, 154
    available transfer capacity, 154
    net transfer capacity, 154
Transient stability margin, 69
Transient voltage instability, 97

Transmission reliability margin, 154
TSSC (Thyristor Switched Series Capacitor), 236

Unified Power Flow Controller (UPFC), 246

V–Q sensitivity, 162
VAR, 6
VAR compensators, 86
Voltage loss of, 58
Voltage collapse, 64
Voltage collapse proximity indicators, 115
Voltage control, 55

Voltage controlled load bus, 205
Voltage dependent loads, 81
Voltage regulation, 26
Voltage security, 93
Voltage security factor, 131
Voltage stability, 64
   bifurcation analysis, 76
   bus participation factor, 163
   a direct indicator, 125
   eigenvalue analysis, 161
   maximum loading point, 74, 159
Voltage stability index, 129
Voltage stability limit, 105, 166
   limiting power angle, 105
Voltage stability margin, 88, 127, 155
Voltage stability performance index, 224
Voltage uncontrolled load bus, 206